D1545666

body movement
PERSPECTIVES IN RESEARCH

body movement
PERSPECTIVES IN RESEARCH

Advisory Editor: Martha Davis
Hunter College

THUS SPEAKS THE BODY.

Attempts toward a personology
from the point of view of
respiration and postures

by

Bjørn Christiansen

ARNO PRESS

A NEW YORK TIMES COMPANY

New York • 1972

Reprint Edition 1972 by Arno Press Inc.

Reprinted by permission of Bjørn Christiansen

Body Movement: Perspectives in Research
ISBN for complete set: 0-405-03140-8
See last pages of this volume for titles.

Manufactured in the United States of America

━━━━━━━━━━━━━

Library of Congress Cataloging in Publication Data

Christiansen, Bjørn.
    Thus speaks the body.

    (Body movement: perspectives in research)
    Reprint of the 1963 ed.
    1. Posture. 2. Breathing. 3. Personality.
I. Title. II. Series.
QP301.C57 1972   153   72-342
ISBN 0-405-03141-6

THUS SPEAKS THE BODY,

Attempts toward a personology
from the point of view of
respiration and postures

by

Bjørn Christiansen

Oslo
Institute for Social Research
1963

Preface

This manuscript is a preliminary draft. It has been written with the aim of bringing together in a fairly systematic way various viewpoints, hypotheses and experimental findings concerning somatic and non-verbal aspects of personality dynamics. It is published at this time with the purpose of stimulating interest and with the hope of inviting discussions and exchange of opinions.

The collection of material started several years ago while I was working as a clinical psychologist at the Institute for Child Psychiatry in Oslo. The Institute, at that time headed by the late Dr. Nic Waal, was a rather unique place in its heavy emphasize on a monistic psychosomatic approach toward mental illness and health. However, also in other respects has such an emphasise been somewhat characteristic of the Norwegian depth-psychological scene, through Reichian influences in general and through the works of the late Dr. Trygve Bråatöy in particular. I became strongly affected by this train of thought myself, but at the same time I saw the danger of the thoughts congregating into an eccentric, exclusive, superclinical orientation. Not only did I feel the need of someone bridging the gap between clinical and more academic ways of conceptualizing the phenomena in question, i.e. the need for an experimental clinical psychology (or a clinical experimental outlook), but also - as a first step - the need of someone spelling out in greater detail the main assumptions and theoretical propositions underlying this "Norwegian school of depth-psychological thinking".

As a fellow of the Foundations' Fund for Research in Psychiatry at the Menninger Foundation in 1960-62, I got the opportunity to continue the work previously started in Oslo. Of particular importance was the unceasing stimulation, encouragement and help given me by Dr. Gardner Murphy and his research department, and the Foundation's excellent library facilities. I am mentioning this latter point specifically, because I hadn't imagined beforehand how much work that have already been published in the field of my concern. Not only did I come across unknown clinical studies but also a number of extremely relevant reports based upon laboratory experiments. Instead of focusing upon a few Norwegian contributions, I gradually became more

and more interested in presenting a comprehensive review of the
literature interlocked with a few personal remarks. However, before
being able to carry through with this plan, I became trapped in a
series of experimental studies on "respiratory pulsations". My
interest in this latter area grew out of the process of writing up
this manuscript. In some respects I would have liked to postpone
the publication of it until I could have incorporated my own
experimental findings. On the other hand, this might have called
for a postponement of several years. The manuscript as it stands
is a preliminary draft, but it also represents, I feel, the completion of a certain stage in my own thinking.

   Most of the manuscript was written during my stay at the
Menninger Foundation, although the final editing and polishing work
was done at the Institute for Social Research in Oslo last winter.

   In summary, the manuscript is meant as an introduction to
an area of research and to a way of conceptualizing psychological
phenomena which I belive not only are facinating and tempting in
their own right, but extremely rich with opportunities for new insights, new understandings, and new approaches both regarding personality dynamics and functioning.

                 February, 1963.

                  B.C.

C O N T E N T

POSTURES AND GESTURES AS PSYCHOLOGICAL PHENOMENA.

Postures and gestures form an important part of human interaction. Investigators of various disciplines have paid attention to these phenomena. Efron (1941) emphasizes the ethnic-cultural aspect, Allport (1937) the personological aspect, and Birdwhistell (1952) the communicative and linguistic aspect of gestural and non-verbal behavior. Allport and Vernon (1933) have demonstrated the style and manner of expressive movements to be consistent for an individual. Reich (1949) has developed an extensive psychodynamic theory of postural configurations. Among psychoanalysts generally, from Freud onwards, specific seemingly chance acts have been looked upon as due to preconscious and unconscious impulses, affects and ideational contents, and defenses against them. It is a fairly wide spread assumtion to-day that no aspect of gestural activity and body postures is left undetermined, and that they are always the the resultant of interacting psychological, biological and social determinants. Our own interest is principally focused on the psychological aspect.

Previous research in this area has suggested a number of fruitful more general viewpoints, but turning to specific behavioral manifestations and interpretations - we are still not very far ahead. In this introductory chapter we want to explore some hypotheses concerning the reflection of personality characteristics in gestures and postures. Specifically, we want to concentrate on the work of George F. Mahl and Felix Deutsch.

1. Mahl's study of gestural behavior in the interview situation.

To what extent is it possible through observations of gestural behavior only to arrive at valid information concerning major personality characteristics? What is the relationship between verbal and non-verbal behavior in a diagnostic interview situation? These problems are dealt with in a recent study by Mahl, Danet and Norton (1959).

As subject were used 18 patients, 9 women and 9 men, seeking help from a psychiatric outpatient clinic. The patients are described as representative of the type of patients seeking help from the

clinic. The behavior studied was that manifested by the patients
during their intake interview. Each interview was recorded on tape
and watched through a one-way mirror by an observer not hearing any
of the verbal exchange. The observer on his part dictated a running
account of the patient's behavior onto a second tape, and also, when
appropriate, noted his "on-the-spot" impressions about the psycho-
pathology and general traits of the patient. Later on, the gesture
record was processed by tabulating all the actions noted for
successive one or two minute intervals. In tabulating the actions
noted the following categories were used:

a) <u>General postural changes</u> - - recrossing of legs, or changing
   trunk position, etc.

b) <u>Communicative gestures</u> - - pointing, shaking head, pounding with
   fist, illustrative gestures, etc.

c) <u>Autistic activity</u> - - playing with jewelry, scratching, rubbing,
   random touching of various parts of body, etc.

After absorbing and digesting as much of the behavioral
data as he could, listening to the gesture tape, studying and counting
specific gestures and actions made during the interview as a whole,
the observer wrote down whatever predictions he felt he could make
about each patient's diagnosis, leading conflicts and character
structure. Finally, the predictions and the "on-the-spot" impressions
recorded on the original gesture tape, were compared with independent
criterion materials. The latter consisted of the verbal content of
the intake interview, and, most importantly, entries in the clinical
case records - the clinical diagnosis, descriptive symptomatology,
and the discharge summary.

Alltogether, 50 major predictions were made. Among these,
43 were found to be clearly confirmed. In some cases, accurate,
precise impressions were found to have arosen spontaneously in the
mind of the observer very soon after the interview started. In one
instance, the observer made the statement at the second minute of
the interview that the patient "Has a homosexual problem and is very
embarrassed about it". This was precisely the focal point of the
interview and of a few subsequent interviews at the clinic. In
another instance, the observer noted also in the second minute,

that this patient must have an authority problem, i. e. fear of
authority. The case description of the patient confirmed the
immediate perception by the observer.

The investigation is not free from methodological shortcomings. The original observer was the person who made the predictions
and also subsequently studied the criterion materials and decided
if there was relevant criterion material available, and if so,
whether the predictions were confirmed or not. Although only un-
interpreted, obviously relevant, verbatim statements in the case
records or the initial interview were used in testing the predictions,
the procedure is not "waterproof" and emphasizes the exploratory
nature of the study.

The main problem confronting us is concerned with 1) what
sort of predictions were made, and 2) on what bases did the observer
make his predictions?

The first and most general principle adhered to was that
any discrete gesture or action which a patient repeats relatively
frequently, or any characteristic posture, are indications of basic
attitudes, drives or conflicts. In illustrating this principle we
may turn to some concrete examples presented by Mahl et al.

"Mrs. Boun (see sketch no. 1) performed the communicative gesture
of turning her palms out-up once every minute and shrugged her
choulders about every three minutes of the interview... Often she
did both at the same time. These acts were regarded as manifestations
of a deeply ingrained attitude of passive helplessness, tainted with
a negativistic, complaining flavour. We also assume this behavior
functions as an appeal to the interviewer. The autistic ring-play,
first with her wedding ring on her left hand and then with an unknown
ring on her right hand was regarded as a symbolic representation
of conflicting feelings over her marriage. Occational scratching
and inspection of her fingernails, which we considered as weapon for
attack, suggested a rather strong undercurrent of hostility no doubt
related to the negativism suggested by her shrugging shoulders. The
fact that she scratched herself produced the prediction that she
predominately turned her aggression onto herself. All of these
characteristic appear in the criteria data. Her diagnosis was
"anxiety reaction in passive-dependent personality". Her descriptive
symptomatology listed: somatic symptoms, weepy spells, depression,
feelings of inferiority and inadequacy. The discharge summary said:
Major concerns are that too many demands are placed upon her by her
marriage and children, and that she must "bottle-up" her anger and
retaliatory impulses felt towards her husband; complains that husband
doesn't help with care of house and children, and never takes her
out. It should be noted that we did not predict the anxiety or the
somatic symptoms of this patient." (pp. 7-8.)

MRS. BOUN

E. HOROWITZ

Another patient, Mr. Bros, is described as follows:

"Mr. Bros (see sketch no. 2) sits centered in the chair, and leans
against the back; he places his legs in an open cross, one ankle
resting on the knee of the other leg. Each elbow rests in its chair
arm and his hands are either held in the pontifical gesture shown
with all fingers accurately apposed, or he holds his glasses in the
balanced composition illustrated - - they are centered between his
hands and the stems are crossed precisely. The patient was an
engraving of symmetry, precise balance, and delicacy. This behavioral
picture was part of the basis for predicting significant compulsive
features in his make up, as well as an emphasis on control over the
expression of hostility. In his case record, his diagnosis included
"compulsive features" and his therapist observed in his summary that
the patient ordinarily dealt with his violent aggressive feeling "by
very rigid control, passivity and compliance". The glasses - activity
of this patient is also of interest. It was unique for this man,
among the patients who wore glasses. The judge equated the removal
of the glasses with a tendency by this patient to use denial as a
defense. When the therapist later closed this man's series of interviews he commented specifically on "strong use of denial" by this
patient. The latter part of the intake interview itself, when the
patient was removing his glasses, dealt actively with the problem
the patient has in "facing certain things" (i. e. problems), to cite
verbatim words used by both the patient and the interviewer."
(pp. 9-10).

In making predictions about aggression, hostility, their
direction, and their control, some more specific assumptions were
adhered to. Shrugging shoulders, making fist, rubbing or wiping
one's nose, interest in teeth and fingernails were considered indices
of a generally hostile attitude. Scratching, was interpreted as
evidence for a tendency to turn the hostility around on the self.
But most important for predicting conflicts over aggression was the
dimension of freedom - restraint in use of the hands and arms.

A male patient with a "violent temper" as a central problem,
did exhibit the following pattern; in gesturing his hands would
suddenly fly out to the full extent of his arms, and just as suddenly
drop back to his body. Another male patient, described as a sociopath
that "attempts to maintain rigid control of his hostile and aggressive
impulses which seem to be occasionally expressed in ill-considered
manner", did show an extreme limitation of variety and amplitude of
hand gestures, either he kneaded, squeezed, or rubbed them heavily
together, or he parted them ever so slightly. A third male patient,
described as a compulsive personality with "a problem of control over
his aggression and hostility with emphasis on the control", did show

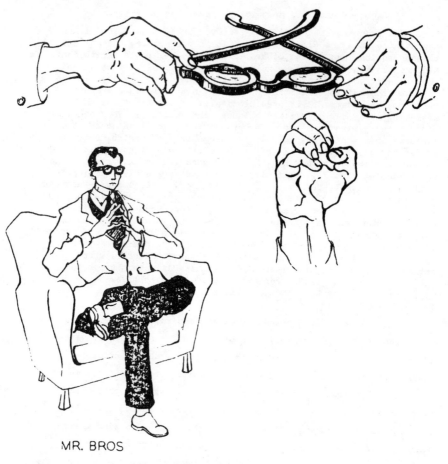

MR. BROS

alternations from moderately free hand activity to that in which his hands were moved together, but most conspicuous was his lowering and raising of his interlaced hand, his frequent energetic pressing of his thumb tips against each other.

As a third principle of interpretation is suggested the rather familiar assumption that a positive relationship exists between the over-all level of anxiety and the frequency of rapid foot movements and general postural shifts.

Furthermore, some specific avtistic gestures were interpreted or follows:

Rubs-touches nose - attitudes of contempt, disgust, negative feelings.

Finger-mouth contact - oral-erotic needs and gratifications.

Rubs self - general erotic stimulation, or tension reduction.

Picking, smoothing, cleaning, etc. - obsessive-compulsive traits.

The last part of the study consisted of comparing the running account of the patients non-verbal behavior (dictated by the observer on one tape) with the verbal interchange actually taking place in the intake interview (recorded on another tape). This was done in order to obtain further understanding of the "meaning" of gestures and movements and to explore the various relationships between verbal and non-verbal activity.

The investigators conclude that there seem to be various types of relationship between what is said in an interview and gestural and body movements. Four types of relationships are suggested:

1) Som actions express the same meaning as the concomitant manifest verbal content. Mrs. Boun described in words the feelings of helplessness expressed in her gesture of turning her palms out-up. Mr. Bros removed his glasses as he and the interviewer actively explored his difficulty in "facing certain things".

2) Some gestures do not appear on the surface to be related to the manifest verbal content, but turn out to anticipate later amplifications of the current content. All the time Mrs. Boun was playing with her wedding ring, she was describing her symptoms, with practically no mention of her husband. Then she verbalized overtly and extensively her complaints that her hus-

band didn't help her and that he contributed to her feelings
of inadequacy. As she did this she ceased completely playing
with her wedding ring. At another time Mrs. Boun was discussing
her feeling of inferiority towards her husband. As she did so,
she momentarily placed her <u>fingers to her mouth.</u> Three minutes
later she stated that her feelings of inferiority actually date
from feelings she had as a child - namely, that she was homely
and not as pretty as her sisters because she then had <u>two front</u>
<u>teeth that protruded badly.</u>

3) <u>Some gestures betray meaning contrary to concurrent verbal content.</u>
A male patient who deftly manipulated a pencil throughout most of
his interview, lost control of the pencil and dropped it to the
floor at one point only - when he was defensively claiming that
his work efficiency was 100 per cent.

4) <u>Some gestural activity and body movements seem directly related</u>
<u>to interaction with the interviewer, with the patient's verbal</u>
<u>content playing a secondary role.</u> After the 8th minute of the
interview Mrs. Boun's gesture of turning her palms out-up dropped
significantly. Towards the end of this initial phase the interviewer manifested considerable emphatic understanding and sympathy
for the patient, events which could be token gratifications for
the need for help sustaining these gestures.

    The authors state that general postural changes of the
patients only made sense when regarded as changes in tension due to
the immediate interaction in the interview. This latter point brings
us over to the works of Deutsch.

2. <u>Deutsch's study of postural behavior in the psychoanalytic situation.</u>

    A posture is the position assumed by the body, as a whole or
by its parts, in order to execute a movement or to maintain an attitude.
Looked upon from such an angle, a posture can always be said to express motivations, motivations on a conscious as well as on an unconscious level.
When a person is invited to lie on a couch and relax, his posture
illustrates not only his ability to relax his muscles, but also, just
as much a pattern of behavior related to the situation and to his
basic personality structure. Given a non-demanding and accepting

situation, e.g. a passive-friendly, relatively timeless psychoanalytic situation, we would expect the situational impact on an individual's posture to be reduced and his postural pattern gradually to express salient personality characteristics. Since such a situation commonly also implies an invitation to express free associations verbally, an exceptional opportunity exists not only through an analysis of verbal utterances and postural observations to arrive at an impression of the individual's basic personality pattern but also to compare and study the relationships between postural and ideational expressions.

Do specific postures accompany specific ideational themes, and do specific ideational themes accompany specific postures in a psychoanalytic situation? Deutsch has attempted to show that this is the case through a number of case studies in which the patient's posture or change of posture during successive analytic hours were recorded concomitant with his verbalized material. Since Deutsch, following psychoanalytic tradition, was sitting behind the patient, only the position of patient's head, arms and legs were systematically observed. Deutsch concentrates on individual cases. Is it possible also to draw conclusions of a more general nature, that is, of invariable relationships between postures and ideational contents generally?

In order to answer this question more systematic investigations would be called for than those reported by Deutsch. However, Deutsch's case studies may represent a valuable first step in answering this question, partly by suggesting hypotheses for later testing, partly by giving us a rough idea of the extent to which individual consistencies are present.

In what follows we will quote from Deutsch case descriptions. We will concentrate on his description of arm and leg postures exclusively and separately. After each series of quotations we will try to generalize his observations in terms of tentative hypotheses for further studies.

Let us start out with his observations of leg postures. The first quotation refers to a twenty-year-old male patient:

1. "Several times when strong homosexual feelings become directed toward the analyst, he bent his legs at the knees, expressing rudimentarily the phantasy of swinging his spread legs backwards from the couch to engage the analyst's head between his legs" (Deutsch, 1947; p. 198).

A twenty-two-year-old female patient:

2. "Becoming aware of a tickling between her legs, she put them apart, and remembered forcing her little brother between her legs and spanking him. She reported a dream about rough housing with boys with fears of being raped, and memories of slapping her brother" (Ibid., p. 201).

The same patient:

3. "Crossing her legs for the first time, she said she had had a 'tiff' with her mother the day before... From that day she kept her legs crossed, the right one over the left. Homosexual phantasies then appeared. She played with her fingers, putting them into the hollow of her cupped hands..." (Ibid., p. 201).

A thirty-one-year-old female patient:

4. "She felt different from other girls, considered herself a failure, ignorant, unimportant, fearful to move, afraid of doing wrong things....she settled rigidly into a posture from which she has so far not deviated: hands resting on her abdomen, legs parallel" (Ibid., p. 203).

A fifty-year-old male patient:

5. "The basic posture. ,was a dorsal position, the right leg over the left...the first break in this...pattern came with memories and phantasies of passive relationship to men: an incident in which he had behaved 'like a sissy', and when a friend treated him like a cripple. In verbalizing these thoughts, he sprawled his legs... Associations followed in which he relived the anxiety of fighting against passive homosexual tendencies..." (Ibid., p. 204).

The same patient:

6. "One day he reported a dream of strong masculine assertiveness. This had been stimulated by meeting a woman who flattered him with her appraisal of his abilities. He felt accepted, superior... He crossed his legs differently, putting the left over the right..." (Ibid., p. 204).

A twenty-seven-year-old male patient:

7. "Only twice during analysis did the patient put his legs in a parallel position: on both occasions he brought dreams which on analysis proved to be phantasies of giving birth to a child." (Ibid., p. 205).

A fourty-three-year-old male patient:

8. "... in one of two positions the legs were held apart, in the other they were crossed. Spreading his legs was associated with his passive (anal) homosexuality, crossing them with his accentuated

masculine protest. A protective hand-on-head posture with crossing of the legs occurred only in the first five months when fears of castration were being analyzed." (Ibid., p. 206).

A forty-year-old male patient:

9. "During three years of analysis the patient (a man with severe spells of anxiety, many phobias and strong submissive, dependent tendencies) almost never changed the position of his legs: the left crossed horizontally over the right. Occasions when he placed his legs parallel or spread them apart were: 1) with phantasies of fearless petting a dog or a horse, 2) with ideas of anal impregnation, 3) when he felt free of fears of castration." (Ibid., p. 207).

A twenty-seven-year-old male patient:

10. "Hostility towards women coincided with putting the right leg over the left; the left over the right when hostile towards men... Any kind of sexual impulses aroused severest feelings of guilt. A temporary compromise was the phantasy of a combined masculinity and feminity, the one represented by an attitude of fistic aggression, the other by spreading his legs apart. During this period he masturbated genitally with heterosexual phantasies, but at the same time with a rubber tube in his anus. From this he regressed to a passive femininity towards both men and women, in which he felt safe. Analysis of the infantile sources of his reactions aroused rebellion expressed in hitting movements of his right arm and transient crossing of his left leg." (Ibid., p. 208).

The same patient:

11. "Homosexual fantasies were accompanied by raising and flexing his knees on his abdomen, presenting his anus." (Ibid., p. 209).

The same patient:

12. "The deterioration of postural behavior during analysis can best be observed in borderline cases when their defensive mechanisms are threatened. A man ... with a severe neurosis bordering on paranoia, ran so rapidly through such a gamut of postures that they were recorded with great difficulty... In states approaching homosexual panic, he assumed as many as twenty-five different positions during one session, exchanging them frequently." (Ibid., p. 209).

A twenty-nine-year-old male patient:

13. "The right leg held down the left one, hiding and protecting the phallus, and preventing lifting the legs as the expression of a yielding to analerotic wishes... Next, he remembered acts of rebellion directed against both his parents. He began to pound the couch with his left hand and, suddenly became aware that his legs were in a parallel position, he remarked, 'the penis is freed.' From that time he never crossed his legs..." (Deutsch 1952; p. 200).

The same patient:

14. "In sexual relationships he would go through a phase of passivity
which he relinquished with great difficulty, and which was followed
by a grandiose, 'crushing', selfish, driving, 'male' attitude, but
actually feeling that he was a 'windbag'. These associations were
accompanied by a feeling of stiffness in his legs, as if he were
afraid of being attacked..." (Ibid., p. 200).

A fourty-eight-year-old male patient:

15. "For a time he became querulous, hostile, aggressive, self-
assertive in the transference. He stated he looked in the daily
newspaper with the thought of finding there a notice of my death.
The posture, left-leg-crossed-over right, which had alternated
with right-over-left, disappeared... He dared (which he hadn't
before) to retract his foreskin during intercourse." (Ibid.,
p. 202).

A young male patient:

16. "During early puberty he developed a peculiar gait and a mannerism
of pointing his toes inward when walking, especially when he felt
he was being observed by girls and when he wanted to display his
masculinity... quick rhythmic up-and-down movements of the feet...
increased... when castration anxiety was evoked in association
with masturbation. Analysis clearly revealed the erotization of
the lower extremities and their unconscious use for the expression
of hostile, aggressive, homosexual impulses." (Ibid., p. 206).

A male patient in the late thirties:

17. "In analysis he lay with... the left leg extended... his right
leg flexed, the sole of the foot resting on the couch. For two
years the left leg never changed its flaccid position... The
passive anal homosexuality was represented by the position of the
right leg." (Ibid., p. 211).

The same patient:

18. "...only after his passive feminine identification with his mother
was partly worked through at the end of the second year did he
suddenly ask with surprise... whether I had noticed that he always
kept his right knee bent upward and his left leg extended and
rotated outward... Thereupon he crossed the left leg over the
right leg for a short while. His associations to this posture
were memories of the continual fear of telling the truth and of
standing up for his ideas and rights. This referred to his feminine whishes represented by the left leg, whereas the right leg represented to him his masculine and independent strivings, the masculine ideal which he could maintain only in phantasy... He assumed
from time to time the crossed leg position with an increasing feeling of self-assertiveness... This led to a short-lived euphoric feeling of having thrown off his 'shackles' accompanied by increased

heterosexual activity. When the elation subsided, his leg
returned to its passive position, from which it was occasionally
aroused by memories of weak rebellious episodes against his
mother's overprotection." (Ibid., p. 211).

As previously stated, Deutsch focuses mainly on case presentations, although in a few places he also tries to formulate more general
principles of postural interpretations. As regards leg postures generally, at one place he states:

19. "In a supine position of the body while the legs are kept parallel
or spread, one or both feet may be flexed, extended, rotated in
or out... The upward and inward position has been analyzed as
representing the effort of resistance against passive tendencies,
and against figures interfering with self-assertiveness. The
outward and downward posture may represent yielding to passivity,
and a change from this position to one with increased muscular
tonus is very often a forerunner of the verbal expression of self-
assertiveness and hostility. In many cases the passive position
means withdrawal as a reaction to unconscious fear of castration
and appears as the preverbal expression of this fear." (Deutsch,
1952; p. 204).

And at another place:

20. "In my experience this type of leg posture (the right leg put
over the left and further flexion of the already crossed leg into
a right angle) has never failed to precede or accompany the verba-
lization of oral phantasies." (Ibid., p. 207).

And at a third place:

21. "...prognosis in analytic therapy becomes possible if the behavior
pattern contains special postures which betray regressive behavior
which apparently cannot be given up. They may be easily recogni-
zed: for example, a hand kept permanently on the mouth, the position of the leg which presents the anus... These regressive
postures are especially ominous when reinforced: both legs are
flexed and raised; or one leg, usually the right, rests flexed
on the other knee, indicating a fusion of anal and oral regression." (Ibid., p. 213).

Reading through Deutsch's statements we get a strong impression that leg postures are intimately connected with sexual, i.e.,
genital, drives and conflicts. Let us briefly indicate the main leg
postures possible in a supine position and which one of these postures
that has been referred to in our quotations. This has to be somewhat
arbitrary, of course, since in many instances detailed information is
lacking.

Survey of main leg postures

(Numbers indicate where referred to in the preceding quotations)

Side View

Above view	Stretched in hip and knee	Moderately bent in hip and knee	Strongly bent in hip and knee
Spread	2,5,8,9,10, 17,19		1,11
Parallel	4,7,9,13,19	13,17,18,20	11,21
Crossed: sharp angle	3,5,6,8,10, 13,15,18		1
Crossed: right angle	9	20,21	

From our survey of Deutsch's descriptions we arrive at the following tentative hypothesis as regards the psychological meaning of leg postures:

1. Genital drives and conflicts related to such drives, are expressed in the tonicity of the legs.

2. Genital conflicts can be warded off in different ways giving rise to enduring postural patterns.

3. Genital conflicts (castration anxiety) in males being warded off in a compensatory masculine pattern corresponds to a leg posture showing relative hypertonicity of extensor and adductor muscles (the legs being kept stretched and crossed (drawn together) with a tendency toward an inward-upward turning of the toes).

4. Genital conflicts (castration anxiety) in males being warded off in a passive feminine pattern corresponds to a leg posture showing relative hypertonicity of rotator and abductor muscles (the legs being kept spread apart and outwardly rotated with a tendency toward an outward-downward turning of the toes).

5. Genital conflicts (vaginal anxiety) in females being warded off
   in a masculine assertive pattern correspond to a leg posture
   showing relative hypertonicity of extensor and adductor muscles.

6. Genital conflicts (vaginal anxiety) in females being warded off
   in a compensatory overfeminine pattern correspond to a leg posture
   showing relative hypertonicity of rotator and adductor muscles.

7. Genital conflicts in males and females being warded off in an analerotic regressive pattern correspond to a leg posture showing hypertonicity of both rotator and extensor-flexor muscles (the legs being
   kept parallel, or parallell flexed and raised).

8. Genital conflicts in males and females being warded off in an
   oral-erotic regressive pattern correspond to a leg posture showing
   hypotonicity of both rotator and extensor-flexor muscles (often
   expressed by one of the legs being outward rotated, flexed, and
   drawn upward so that it crosses the other horizontally over the
   knee).

We want to reemphasize that the above hypotheses have to
be considered as extremely tentative ones. In later chapters we will
discuss these hypotheses further from a theoretical, developmental
point of view as well as from a perceptual and experimental viewpoint.

- - - - - -

Let us now turn to Deutsch's arm posture observations. A
male patient of twenty:

1. "The patient kept his arms rigid and close to the body, which proved
   to represent an attempt to suppress thoughts of masturbation which
   soon afterwards became verbalized. When incestuous wishes towards
   his mother or sister came to expression in the analysis, he began
   holding his hands on his abdomen, above his genitals. As passive
   homosexual cravings emerged, he lay with his arms in a position
   resembling a sleeping baby. Incipient rebellion against his passive
   homosexuality was marked by a change in which one arm remained
   in the infantile position and the other moved downward, with the
   addition that in reviving feelings of competitiveness against his
   father, he put his hand in his trouser pocket." (Deutsch, 1947;
   p. 198).

A twenty-two-year-old female patient:

2. "She clenched her fists when she recalled her mother's anger with
   her, calling herself an unhappy, abnormal person who never got what
   she wanted... She put her hands over her chest and held her fingers
   crossed when feeling as if a snake were creeping on her... Moving
   her hands to her head as if protecting it, she had thoughts of
   'sins' which led to confession of masturbation... holding her hands

under her neck when fearful of being punished for masturbation; her right hand was lifted and her left hand held protectively over her head when she was angry with men. Her left hand was usually lifted when she was in a rage with her mother. Both arms were lifted when she felt hostile with both parents. Both arms were stretched backwards when longing for approval." (Ibid., p. 201).

A fourty-year-old female patient:

3. "Masturbatory feelings of guilt were often accompanied by concealment of the hands under the neck. As a rule, the head was covered with one arm or hand until late in the analysis when she was relieved of feelings of guilt centering about her mother." (Ibid., p. 202).

A twenty-three-year-old female patient:

4. "She felt... fearful to move, afraid of doing the wrong thing... Occasional brief movements of her injured hand towards her head appeared when she related masturbatory practices, and when she touched upon hostile impulses towards her brother; otherwise, whatever her emotions, analysis has so far not altered the rigid, restrained, basic posture (i.e. hands resting on her abdomen)." (Ibid., p. 203).

A fifty-year-old male patient:

5. "The basic posture of a fifty-year-old obsessional neurotic man with a climacteric depression was a dorsal position, the arms on his abdomen... in verbalizing these thoughts (phantasies of passive relationships to men), he spread out his partially flexed arms in a babylike posture... (When later showing masculine assertiveness) he kept his arms rigidly extended." (Ibid., p. 204).

A twenty-seven-year-old male patient:

6. "Passive dependence conflicting with deep resentment against his parents made him a timid, cautious but stubborn personality... During the first three months he held his arms and hands together in a stiff defensive position on his chest. Later, he made specific movements of his arms suggestive of protecting himself against the punishment he feared or as if to attack the menacing figures. He expressed need for protection by holding both arms over his head. When rebellious, he lifted one arm, the other remaining as a safety device over his heart... In the tenth month of his analysis, passive homosexual phantasies became conscious... He became silent and held his arms parallel to his body... Relinquishing his feminine identification and expressing his hostility towards his mother freed him of his dependence which found representation in putting one or both hands into the pocket of his coat." (Ibid., p. 205).

A fourty-three-year-old male patient:

7. "On the analytic couch, he regularly regressed to a passive infantile state, frequently daydreaming... the arms were held over the abdomen or chest... (later after castration fear had been analyzed) the defensive position of the arms was abandoned and the arms lay extended on the couch." (Ibid., p. 206).

A fourty-year-old male patient:

8. "He (a submissive, passive dependent patient) sought treatment for severe spells of anxiety and many phobias. During the first half year of treatment he lay motionless, either with his hands, finger spread, lying on his belly, or his hands held uplifted. As the analysis progressed these postures began to alternate with more aggressive ones... At the peak of reliving his competition with his father and his brother, he held his arms in a position of boxing. Newly acquired selfconfidence coincided with putting his right hand (the left only twice) in the pocket of his trousers, the other in an attitude ready to strike." (Ibid., p. 207).

A twenty-seven-year-old male patient:

9. "When feeling offended or unjustly treated, this patient crossed his arms over the lower abdomen... In the first four months... the predominant position being...arms held in a defensive attitude protecting the head. The protective gestures disappeared almost completely after a few months of analysis... (When later passive yieldings appeared) the arms (were) resting over the chest, or uplifted and resting on the pillow or either side of the head. In the last five months of observation, he began a movement of his left arm, bending it upward at the elbow with a quick rotating movement which seemed to presage an aggressive assertiveness." (Ibid., p. 209).

A twenty-nine-year-old male patient:

10. "In the third year of analysis the trend of his associating was chiefly anal with rebellious aggressive phantasies directed against his father. He became aware that he held his hands over his belly (finger clasped) as if he wanted to prevent its protrusion, and that he did not dare to put them down because he was afraid that if he did other phantasies would come up. He relaxed his hands... remembered act of rebellion directed against both his parents." (Deutsch, 1952; p. 200).

A young male patient:

11. "Postures like folding the arms, clasping the fingers, the foetal position, may be interpreted as attempts to hold the 'parts' of the ego together and to diminish body surface. Movements of the limbs away from the body are rudimentary threats of loss which are counteracted by the antagonists." (Ibid., p. 208).

A male patient in the late thirties:

12. "A man in his late thirties sought treatment for chronic mild depression and feelings of inadequacy... In analysis he lay with his head turned slightly to the left, the left hand on his chest... His right hand grasping the pillow. The position of the left hand was revealed to be the expression of his passive oral dependence on his mother, which included also the turn of his head to the left... His need to cling to father for support was expressed by the right

arm and hand. Deviation from this basic posture appeared in the course of the analysis according to the alleviation of his dependence on father." (Ibid., p. 211).

The posture of the arms will depend upon the tensional state of a number of muscle groups, of the shoulder, the upper arm, the fore-arm and the hand. The posture of the hand itself may show great varia-tion. Deutsch (1949) distinguishes between three main hand postures: 1) cupped hands, the right hand held over the left or the left over the right, 2) flat hand position, the hands not touching each other but held more or less close together, and 3) clasped hands, the fingers being held interlocked and interlaced.

In a study by Grieg, et al., (1957) the following seven hand postures are differentiated: 1) clenched fist with free thumb, 2) clenched fist with covered thumb, 3) claw hand, 4) delivery hand, 5) pill-rolling hand, 6) stiffly-stand-out hand, and 7) normally relaxed hand. Deutsch too mentions the many variations possible in such a relatively small detail as the thumb position in a clasped hand posture: the thumb may be covered by the fist, one thumb may be capped over the other or touching the other only with the uppermost joint.

The arm postures referred to in our quotations represent far from all the arm postures possible. The number of times each of the postures referred to, is mentioned, are relatively few. However, reading through Deutsch's observations we get the general impression that arm postures are intimately related to such personality variables as asser-tiveness, dominance, submission, aggression, restraint, longing, and contact-seeking. At this point it is interesting to recall that Mahl et al. for predicting conflicts over aggression found the dimension of freedom - restraint in the use of the hands and arms - to be most use-ful. The following hypotheses are suggested by Deutsch's observations:

Posture	Referred to under No.	Corresponding psychological attitude
Hands uplifted-grasping pillow	8,9, 12	Craving for dependency and support
Arms lifted-boxing position	2,6,8	Anger, aggressive impulses
Hand covering or held over head	2,3,4,6,9	Craving for protection, guilt over hostility

Posture	Referred to under No.	Corresponding psychological attitude
Hands kept under neck or head	2,3,4	Suppression of or guilt over masturbation
Arms mildly flexed-helt in pocket	1,6,8	"Self-confident" assertiveness
Arms stretched backward, rotated	2,5	Longing for affection
Arms rigidly extended	5	Compensatory self-assertiveness
Arms kept close, i.e. parallel to body	6	Craving for submission
Arms held folded over chest	11	Protective retentiveness, fear of letting go (anger?)
Arms crossed or clasped over chest	1,2,6,7,9,11, 12	Passive receptive oral yielding
Hands clasped over abdomen	4,5,7,8,10,11	Fearful restraint of anal expulsive impulses (fear of loss)
Hands held covering genitals	1,11	Conflictual genital wishes

Again we would have to emphasize the tentative and exploratory nature of the hypotheses stated. As regards hand postures specifically, we would like to mention in passing that a stiffly-stand-out hand in one case at least, was accompanied by a fearful, anxious state of mind, and clenched fists, by verbal expressions of hostility and self-assertiveness, although we don't have enough empirical observations to refer to in this area to suggest any more specific hypotheses.

Given the viewpoint that an individual's posture in a non-demanding and maximally accepting situation will express basic personality traits and drive constellations, we should expect not only that the individual's postural pattern will remain relatively constant over time but also that more basic personality changes brought about by psychiatric treatment, would show up in postural changes.

As regards the first question on postural consistencies, Deutsch (1952) reports that in twenty-three cases with forty-three interruptions of the analysis, interruptions usually lasting from two to four months, comparisons of the patients' basic posture in the last hour with their posture in the first hour after the interruption, brought out that with only one exception their postures were identical. The identity of the postural patterns after interruptions of considerable duration shows nothing other than that the patients' psychodynamic balance had remained unchanged, Deutsch maintains.

As regards the second question raised, the following quotation may serve as an illustration of Deutsch's view:

"It is not the variability of postures, nor the number of changing postures which are of importance for the evaluation of therapeutic progress, but their flexibility during the dynamic process...(A) poor prognosis is made from a basic posture which has remained unchanged after several interruptions. That proved true in six of the cases studied. One of them, a passive man and a latent homosexual with schizophrenic traits, whose basic posture was checked after four intervals of from forty-eight to one hundred sixteen days, retained the same basic posture. After remarkable improvement of his personality and in his professional life, the analysis was terminated in the fourth year. However, a relapse occured and the dubious prognosis proved to be justified, as it was in the other five cases with a similar inflexibility of the basic posture." (p. 214).

The aim of psychoanalysis and psychotherapy is to change the psychodynamic balance of patients, to change the equilibrium among instinctual forces and impulse-defence configurations. To the extent these forces, equilibria and configurations have known somatic equivalents, observations of these equivalents (as they eventually are expressed in basic postural patterns) may consequently serve as an extremely valuable method for the evaluation of analytic therapy as well as for diagnostic purposes generally. In fact, Deutsch suggests that we are here faced with a postural method far better than the psychological ones commonly used.

3. Some concluding remarks concerning postural and ideational behavior.

The task of interpreting how the body silently speaks is far from an easy one. So far, much more research has been directed toward the interpretation of Rorschach protocols than toward the understanding of postural behavior, and consequently, for most psychologists today verbal behavior is more real and to a certain extent considered more significant than other types of behavior.

In stating that the task of interpreting postural behavior
is far from an easy one, we don't imply that it is either easier or
more difficult than interpreting verbal behavior. In fact, a number
of problems seem quite similar in the two areas. In both areas we are
confronted, for instance, with the problem of latent versus manifest
behavior, with situational influences being superimposed upon more
basic patterns, with superficial and deep tensional states.

In the same way as a psychologist feels safer in his diagnostic evaluation the longer he has talked with a given patient, and the
more psychological tests he has administered, tapping different traits
and dispositions, so would a posturologist (or a psychologist with
equal training in understanding postural behavior as that usually given
for the understanding of verbal behavior) certainly feels more safe the
longer he has observed a given patient and the more postural tests he
has administered tapping different traits and dispositions. He would
certainly be very reluctant to make too much out of a single or a few
very short-time observations of an individual's postural characteristics
as he is lying on a couch. He would be aware that situational factors
as apprehension related to the unfamiliar situation, the fact that the
patient has eaten something poisonous the day before and has pain in
his stomach, the fact that he has just quarreled with his wife and feels
guilty, the fact that he has resently stumbled and strained his leg, etc.,
might easily color the postural pattern being expressed and hide more
basic properties from direct observation. Also in dealing with more
basic, characteristic postural patterns would have he to show restraint
and cautiousness in his interpretations. We have already indirectly
touched upon the possibility that certain postural configurations may
include ambivalences and conflictual sets, and that certain postures may
be superimposed upon others, probably genetically earlier ones. We
mentioned in one of our quotations, for instance, that an upward, inward
position of the feet with legs spread out may represent an effort of
resistance against passive yieldings, an attempt toward the warding off
of passive-feminine impulses by a hyper-masculine protest, the passive-
feminine impulses and passive yieldings in turn probably representing
an earlier and more primary defense stratagem of warding off genital
anxiety and conflicts. Besides the problem of dissolving and dissipating

hierarchical arrangements of postural patterns stemming from the same
tensional muscles (for instance, genital conflicts or anal conflicts),
he would also have to take account of the possibility of several unresolved tensional nuclei existing in the same individual and that
their postural defenses and expressions may overlap to a considerable
extent. For instance, although we would think that somewhat different
muscle groups in the legs probably are involved in the postural defense
and dissolution of anal-expulsive and genital-intrusive impulses, we
would still expect the structuring of leg tensions in the case of genital conflicts, to be strongly influenced by whether unresolved anal
conflicts exist or not. In principle we would think the same dynamics
apply to most body parts. But not only would our posturologist have
to take account of antecedent conflicts influencing the postural solution of later ones, he would also have to consider the possible effect
of cohesive and synthesizing forces of a maturational nature acting on
initially relatively discrete postural solutions. He would, in short,
have to base his interpretations on a fairly comprehensive structural-
developmental personality theory. In this as well as in all other
respects mentioned, he would not be worse off than the psychological
clinician working with verbal responses and the most commonly used
psychological inventories. Here too, a theoretical model of one sort
or another is a _sine qua non_ for personality description and interpretation. But our friend, the posturologist-psychologist, would be, in 'the
long run we feel, perhaps in a better position since his personality
model most probably more easily would take account of verbal utterances,
content and style-wise, than a model primarily anchored in verbal
behavior could be able to account for postural phenomena. On the other
side, the expression _in the long run_ would have to be emphasized since
the research data existing today - in spite of their discursive quality
and quantity in the field of personality dynamics - give the verbally
oriented psychodiagnostician and psychotherapist a considerable lead in
knowledge and tradition. Of course, to present the matter as a question
of either-or, as a question of which viewpoint is the most profitable
one, is to dramatize a conjecture that doesn't really exist. Faced
with the study of such an enormous complexity as the human personality,
different approaches are needed in furthering our understanding and

insight, and all approaches, proved to be valid, would complement and enrich each other. To the extent postural viewpoints further our understanding they are neither more nor less valuable or fruitful than any other viewpoints. This does not mean, however, that one is bound to think that all angles offer the same possibility of a break through into unknown territories and into systematic research possibilities of crucial psychological problems.

Following this somewhat discursive introduction and our final rather personal confession, let us turn to the respiration as a postural and psychological phenomenon.

RESPIRATION AS A POSTURAL AND PSYCHOLOGICAL PHENOMENON

We mentioned in the last chapter the possibility of using postural manifestations as a source of psychodiagnostic information. An individual's respiratory pattern will be strongly influenced by his postural configurations. Thus it is not at all surprising that the respiratory pattern has been suggested to offer psychodiagnostic insights. Some quotations may serve as an illustration of this point. Kempf (1930) writes:

"We are all well aware that we are often able to read character and personal attitude and affective disturbances by the manner in which people breathe and carry the thorax in certain situations. That is to say, the postural tensions of the thoracico-diaphragmatic abdominal musculature and the manner of breathing often reveal the courage, timidity, egotism, humility, frankness, arrogance, pride, subjugation, friendliness, shyness, depression, or gaiety of the individual in the situation." (p. 184)

Braatöy (1954) writes:

"In relation to the patient, the problems of respiration focus the attention on a vital function which is always modified, restricted, or blocked in states of attention, watching, and in suppressed and repressed emotions. For this reason, the patient's breathing is the best index in gauging his emotional state." (p. 179)

Lowen (1958) presents somewhat similar thoughts:

"...respiratory movements occupy an important position in bioenergetic analysis. We look to see if the chest is expanded and held rigid or soft and relaxed. A blown up chest is the invariable concommitant of a blown up ego. It reminds one of the fable of the frog who attempted to blow himself up to the size of a bull. On the other hand, a soft chest, although related to more feeling, is not necessarily a sign of health. It is found in certain impulsive character types who have a pregenital structure. What we look for is a relaxed structure in which the respiratory movements show the unity of chest, diaphragm and abdomen in inspiration and expiration." (p. 90-91)

Discussing the normal, healthy respiratory pattern, Braatöy (1947) focuses on the same unity:

"Recorded with pneumographic curves...the respiration curve of normal subjects in a relaxed, supine position will run as a "sinus curve." The respiration curve runs as a recumbent S. From the peak of inspiration the curve runs smooth and well rounded by the

downhill expiratory curve in order to oscillate equally smooth
in a new upward turn, and so the curve continues with steady, even
crescentric oscillations above and below its middle course...normal
healthy people breathe in this way, and if one simultaneously
reads of their abdominal respiration, one finds here a similar
curve parallell to the chest respiration. By direct observation
it is difficult to take hold of this in itself fairly peculiar
play of the chest cavity expanding smoothly in <u>all</u> directions at
the same time; that consequently the diaphragm pushes the stomach
down and the abdominal wall forward at the very same time as the
chest expands. If one first has become alive to this phenomenon
one has the basis for recognizing how often people <u>don't</u> breathe
in this way. Either the chest or the abdomen expands or the
abdomen falls out of time, first the one and so the other. Or
the individual respiration waves are uneven in such a way that
the regular respiratory rhythm gets displaced to an irregular
arrhythmic function." (p. 230-231)

Both Braatöy and Lowen suggest a sort of "base line" or
"healthy respiratory pattern" in relation to which individual's
breathing patterns may be compared and evaluated. A number of possible
deviations were alluded to: a fixated, blown up chest, a lack of
synchronization between the movements of the chest and the abdomen,
and a partial or total arrest of abdominal movements.

As was the case with Deutsch's conception of basic postural
patterns, so has respiratory recordings too been considered a valuable
potential tool in ascertaining therapeutic progress and prognosis.
Describing their experiences with spirographic recordings of patients
under psychiatric treatment, Sutherland, Wolf, and Kennedy (1938)
state:

"...an important fact is that as a patient's mental stage changes,
so too did his accompanying spirogram. As a patient improved
clinically, his respiratory 'fingerprint' or spiroprint approached
the normal. Conversely, if his mental state became worse, his
spirometric tracing showed it; i.e., became more irregular and
bizarre. In fact, from a glance at successive records the progress
of a case could be followed without actually seeing the patient.
Clinical examinations confirm this. Here then, is an objective
gauge of a patient's mental state independent of subjective evaluation by both patient and doctor; a convenient easy method by which
to follow a patient's progress in a simple and timesaving manner
and one further that has the advantage of eliminating the personality
of the physician... It is likely that the pattern of respiration
is in a sense as true an expression of thalamic and medullary
activity as the Berger wave is of cortical rhythm... In our spirograms every neurotic or psychotic patient produced at least some
slight but detectable departure from normal respiration."
(p. 103)

Following these very optimistic and encouraging statements
let us turn to the most basic question involved in the discussion, to
what extent do we find consistent individual differences in breathing
patterns?

Again we may quote from Sutherland, Wolf and Kennedy's (1938)
article. They state:

> "It seems remarkable . . . that records taken three to twenty-
> eight days apart, can reproduce each other with such precision.
> Yet this is the fact in each case. If duplicate records taken
> days apart in several hundred cases are shuffled, they can be
> paired quite easily. Consequently, we have come to call these
> spirometric pictures 'fingerprint' records of our patients. These
> records could not be duplicated by any conscious process."
> (p. 103)

Very much the same viewpoint is presented in a paper by
Alexander and Saul (1940):

> "Comparing a series of curves of one individual with a series ob-
> tained from another individual, two facts are immediately apparent.
> The first is that the curve is rather typical of the individual,
> like his handwriting. In other words, there are characteristic
> differences between the respiratory patterns of different indivi-
> duals, just as their handwritings are different. In our series
> no two individuals have yielded identical respiratory patterns.
> The second fact is the constancy of any individual's respiratory
> tracing. In about three-quarters of the Institute cases, despite
> variations in details, the major features remained characteristic
> of the individual over long periods (at least three years, the
> duration of our studies). Those which showed considerable variabi-
> lity still retained recognizable individuality. Experiments showed
> imitation of another's spirogram to be extremely difficult."
> (pp. 115-116)

The comparison of respiratory tracings to "handwriting" and
"fingerprints" certainly suggests a high degree of intra and interindividual consistency. Can respiratory patterns give rise to inferences
in the same way as those claimed in graphology and palmistry? To this
question we will turn later on. In what follows we would like to dig
a little deeper into the question of individual differences in respiration.

Individual differences in breathing

How people breathe is of course an empirical question. One
way to get some information about this topic is to consult textbooks

in anatomy and physiology. Checking various textbooks one finds that
somewhat different conceptions exist as to what normal respiratory
movements are like. This does not necessarily imply that one conception
is more correct than another, but that various textbook writers may have
a tendency to overgeneralize their own personal observations.

Although respiration may be brought about in different ways
in different individuals - as we are going to show in this chapter -
it would principally always consist of two phases, inspiration and expiration; inspiration being caused by an increase in the thoracic
cavity - thereby forcing air into the lungs - and, expiration, by a
decrease in the thoracic cavity - thereby forcing air out of the lungs.

In the 1942 edition of Gray's Anatomy of the Human Body we
are told the following about inspiration:

> "The diaphragm is the principal muscle of inspiration. During
> inspiration the lowest ribs are fixed, and from these and the
> crura the muscular fibers contract and draw downward and forward
> the central tendon. In this movement the curvature of the
> diaphragm is scarcely altered, the dome moving downward nearly
> parallel to its original position and pushing before it the
> abdominal viscera. The descent of the abdominal viscera is
> permitted by the elasticity of the abdominal wall, but the limit
> of this is soon reached. The central tendon applied to the abdo
> minal viscera then becomes a fixed point for the action of the
> diaphragm, the effect of which is to elevate the lower ribs and
> through them to push forward the body of the sternum and the
> upper ribs." (p. 403)

What is here described is the assumed muscular actions behind
what we may call a predominant or exclusive abdominal inspiration.
Only after the abdominal excursion, will the expansion of the chest take
place. In other words, we should expect a certain time lag between
the abdominal and the thoracic inspiratory movements. One may ask,
what about the role of the intercostales muscles in the elevation of
the chest. On this question we are told:

> "The intercostales interni and externi have probably no action
> in moving the ribs. They contract simultaneously and form strong
> elastic supports which prevents the intercostal spaces being
> pushed out or drawn in during respiration. The anterior portions
> of the intercostales interni probably have an additional function
> in keeping the sternocostal and interchondral joint surfaces in
> opposition, the posterior parts of the intercostales externi
> performing a similar function for the costovertebral articula-
> tions." (p. 403)

In the 1948 edition of the same book, the conceptions of the muscles involved in inspiration is somewhat different. Under the headline "Quiet Inspiration", we learn that:

"The diaphragm contracts, increasing the vertical diameter of the thoracic cavity. The first and the second ribs remain fixed by the inertia and resistance of the cervical structures, <u>and the remaining ribs, except the last two, are brought upward toward them by the contraction of the intercostales externi.</u>" (p. 391)

What is here described, is the central feature of what we may call an <u>unisonal inspiration.</u>

Under the headline "Deep Respiration" in the 1942 edition of the same textbook we are told that:

"In deep inspiration the shoulders and the vertebral borders of the scapula are fixed and the limb muscles, Trapezius, Serratus anterior, Pectorales, and Latisimus dorsi, are called into play. The Scaleni are in strong action, and the Sternocleidomastoidei also assist when the head is fixed by drawing up the sternum and by fixing the clavicles. The first rib is therefore no longer stationary ... the umbilicus is drawn upward and backward . . . The thoracic curve of the vertebral column is partially straightened, and the whole column, above the lower lumbar vertebrae, drawn backward." (p. 404)

The last description corresponds to what we may call a <u>costal inspiration.</u>

The last quotation introduces an important point, the close interaction between the thorax and the vertebral column, the fact that flexion and extension of the spine corresponds to a depression and raising of the ribs independently of any motion in the costo-sternal and interchondral joints. Thus the back muscles and their state of tonicity may be considered to have important respiratory consequences. The same may be said of the shoulder girdle. The movement of the thorax has repercussion on the position of the shoulders, and the latter, important effects on the thoracic movements. But not only may the shoulder girdle and the back muscles be considered an integral part of the respiratory apparatus, so may also the pelvic girdle and the abdominal muscles. All the abdominal and back muscles influence the position of the pelvic girdle, and are themselves influenced by this structure. The descending of the diaphragm during inspiration increases the intra-abdominal pressure not only in an anterior direction, but also downward toward the pelvic floor. Thus the tonicity of the muscles

stretched across the base of the pelvic will also have a secondary
respiratory function, and so will the muscles in the front of the
neck in the opposite end of the trunk - by tensing during inspiration these muscles prevent the inward suction of the soft parts of
the cervical facia, a suction which otherwise might compress the
great neck vessels and the apices of the lungs. Thus not only the
chest, the back, the shoulders, the abdomen, the pelvic floor, but
the neck too, represents regions in which respiratory movements might
be found. In fact, the human body being built the way it is, one may
start to wonder <u>if not all bodily musculature participates to a greater</u>
<u>or lesser extent in respiratory movements.</u>

Principally, we may differentiate between three ways by which
the increase in the volume of the thoracic cavity can be brought about,
namely, (a) by the descent of the diaphragm from contraction of its
muscle, (b) by the expansion of the thoracic wall through the action
of various muscles on the ribs, the sternum and vertebral column, and
(c) by a combination of these two ways. Parallel to these ways a
division might be drawn between abdominal, costal and unisonal inspiration according to which body region show the greatest and most pronounced
changes in circumference.[1]

The fact that different textbooks in anatomy present different
pictures as to what constitutes normal respiratory movements, strongly
suggests the existence of marked individual differences in this area.

So far we have dealt mainly with the inspiratory aspect of
respiration. Turning to the expiratory aspect, we here too immediately
find differences of opinion as to which muscles normally are involved.

Very much parallel to the distinction made between diaphragmatic and costal inspiratory movements, we have (a) the contraction
of the abdominal muscles increasing the abdominal pressure and thereby
forcing the diaphragm upward, and (b) the action of certain muscles
of the ribs and vertebral column retracting the thoracic wall. But
in addition to this distinction, we have another one -- and perhaps
a much more important one -- the amount of muscular action really

---

[1] Or one may go one step further, as for instance Halverson (1941)
has done, and differentiate between six types of respiration:
exclusive costal, predominantly costal, unisonal, predominantly
abdominal, exclusive abdominal, and paradoxical, the last one
referring to a complete dyssynchronization between thoracic and
abdominal movements.

being invested in the expiratory movement.

In the 1942 edition of Gray's <u>Anatomy</u> under "Quiet Experiation" we are told that: "Expiration is effected... by the action of the abdominal muscles, which push back the viscera displaced downward by the diaphragm." (p. 404). In the 1948 edition of the same textbook; that:

"The normal resting position of the thorax . . . is restored without muscular efforts after a quiet inspiration by the recoil of
the structures which were displaced by the inspiratory act. The
displacement of the anterior abdominal wall is overcome by the
tonus of the abdominal muscles. The ribs are restored from their
displacement by the elasticity of the ligaments and cartilages
which hold them in place. The extensive network of elastic fibers
which permeates the pulmonary tissue retracts the diaphragm, helps
to draw the latter back up by its elastic recoil." (p. 391).

Thus in the one case, expiration is considered normally a passive movement, in the other, at least in part, an active one.

We may stop at this point and inquire into how great excursions and muscular contractions do actually occur at different body regions during respiration.

According to Gray's <u>Anatomy</u> the manubrium sterni moves on the average 30 mm. in an upward and 14 mm. in a forward direction during deepest possible respiration, while the width of the subcostal angle, at a level 30 mm. below the articulation between the body of the sternum and the xiphoid process, is increased by 26 mm. and the umbilcus drawn upward for a distance of 13 mm.

One of the most comprehensive studies made so far of breathing movements, is a study by Wade (1954). In addition to measuring his subjects ventilation (spirometrically), Wade obtained independent measures of changes in chest circumference at the lower end of the sternum, of the vertical movement of the thoracic cage at the level of the sternoxiphisternal joint, and of the vertical movement of the central tendon of the diaphragm (fluoroscopically), with his subjects in both an erect and a supine position.

Below is presented Wade's main results. In figuring out the diaphragmatic movement care was taken to correct for vertical movement of the chest wall, a lifting of the wall implying a heightening of the anterior attachment of the diaphragm. By making such corrections Wade arrives at measures indicating much larger diaphragmatic excursions than what has previously been supposed to be the case. While Gray's

Wade's measures of ventilations (ml.) and respiratory movements (mm.)
in normal subjects (N = 10).

		Erect posture		Supine posture	
		M	SD	M	SD
a.	Inspiratory capacity	3170	450	3870	540
b.	Expiratory reserve	1680	520	930	320
c.	Vital capacity (a + b)	4860	840	4800	770
d.	Tidal volume	799	210	758	161
e.	Tidal-total ratio (d:c)	16%	–	15%	–
f.	Complemental chest mov.	59	15	60	16
g.	Reserve chest mov.	15	6	18	5
h.	Total chest excur. (f+g)	74	18	77	17
i.	Tidal chest mov.	12	4	7	2
j.	Tidal-total ratio (i:h)	16%	–	9%	–
k.	Complemented diaphr. mov.	64	10	80	13
l.	Reserve diaphm. mov.	39	16	19	7
m.	Total diaphr. excur. (k+l)	103	22	99	16
n.	Tidal diaphr. mov.	17	3	8	3
o.	Tidal-total ratio (n:m)	16%	–	18%	–
p.	Total insp. vertic. chest lift.	26	13	18	9
q.	Total exp. vertic. chest low.	4	7	2	2
r.	Tidal resp. vert. chest mov.	none	–	none	–

Anatomy talks about an average excursion of about 12 and 29 mm. during
quiet and forced respiration, Wade suggests an average excursion of
about 17 and 100 mm. On very much the same line he suggests "that in
a full vital capacity about one-quarter of the ventilation is due to
chest expansion and three quarters to diaphragmatic movement." (p. 204).
As shown in the table of results, in ordinary tidal respiration an
individual utilizes on the average only around one-sixth of his ventilation capacity, his chest-circumference-change-capacity, and his
diaphragmatic movement capacity. This last observation clearly indicates an individual's large opportunity to pattern his respiration
according to something else than his basic need for oxygen consumption.
The large standard deviation found in relation to all the variables
studied give further evidence to the fact that great individual differences exit in respiratory pattern.

Turning to the question of muscular action potentials involved in respiration, Haavardsholm (1946) has presented data from a small sample of normal subjects, showing the average EMG potentials from above the xiphoid process to equal 3 and 6 microvolts under expiration and inspiration respectively. The voltage found during inspiration showed a range from 3 to 15 microvolts, and during expiration, a range from 0 to 6 microvolts. In this study too great individual differences were found. For instance, in 20% of the recordings the expiratory voltage was found to equal the inspiratory one, while in another 20%, the expiratory voltage was found to be approximately zero. The average inspiratory-expiratory ratio in action potentials was found to be approximately 2.0.

Haavardsholm in a small study also tries to answer the question whether a correspondence exists between action potentials from the xiphoid process and circumference changes of the abdomen. He finds a gross positive relationship, the average pneumographic abdominal amplitude being 8.5 mm., the average voltage increase being 4 microvolts, and each one mm. body expansion corresponding grossly to an action potential increase of around 0.45 microvolts. When inviting his subjects to breathe more forcefully and more deeply, he found a marked increase in EMG potentials, the average during inspiration reaching 10 microvolts, while invitations to let go and relax as deeply as possible during expiration, was found not to reduce the voltage (if any) previously recorded during quiet respiration. Consequently, his findings indicate that action potentials easily may be voluntarily increased above the level found in quiet breathing, but scarecely descreased voluntarily below this level.

Another point emphasized by Haavardsholm is that forced thoracic breathing does not give reduced but increased action potentials from above the xiphoid process. This latter observation corresponds closely to Wade's (1954) fluoroscopic findings, namely that diaphragmatic movements ordinarily are outside voluntary control and that attempts to cease them (directly or by contraction of the abdominal wall) usually are accompanied by vertical upward movements of the chest wall paralleling the downward movement of the diaphragm. Haavardsholm alludes to the same explanation, and states that small inspiratory action potential at the midriff most probably not so much indicates costal predominance of breathing (as he had previously assumed), as restricted, inhibited and careful breathing movements generally.

We want to stress this latter point because it brings us back to some observations quoted in the beginning of this chapter, to Sutherland, Wolf and Kennedy's observation that their patient's respiratory records "could not be duplicated by any conscious process," and to Alexander and Saul's remarks that "experiments showed imitation of another's spirogram to be extremely difficult." That some people are much more able to control voluntarily their respiratory pattern than others seems likely in view of, for instance, Christie's (1936) and Braatöy's (1954) case description of mental patients being completely unable to obey invitations to modify their respiratory rate or depth or their predominance of costal or abdominal movements.

By this remark we have, of course, taken the first step into the next area of concern to us, the psychological significanse of respiratory patterns. In order to approach this area let us start out with a more general question: What determines differences in respiratory patterns? Immediately we are confronted with two types of differences, between situational ones and interindividual ones. Since, as alluded to above, consistent individual differences seem to be present, we will start out with this latter question.

Casual factors behind inter-individual differences in breathing.

A widespread belief is that individual differences in breathing is caused by structural anatomical differences. To exemplify this viewpoint, and in order to discuss individual differences generally, let us in the following concentrate on sex differences exclusively.

In Piersol's Human Anatomy (1907) we are told:

"In the female the upper portion of the thorax is less compressed from before backward and is more capacious than in the male. The upper aperture is larger and the range of movement between the upper ribs and the sternum and vertebrae is greater. These circumstances account both for the fullness of the upper portion of the chest in the female and for the characters of the respiratory movements, which is known as thoracic; while that of the male, in which the lower ribs and abdominal walls move more freely, is known as the abdominal type of respiration". (p. 168)

Piersol's anatomical explanation of sex differences in breathing represents a traditional viewpoint going back to studies made in the

middle of the last century. Although Piersol considers individual
variations very much structural-anatomical and hereditary in origin,
this does not imply that environmental factors are ruled out completely,
e.g. Piersol differentiates between three types of chest postures,
keeled chest or "pigeon breast", flat chest, and barrel chest, each
pointing to specific genetic events; i.e. somatic disease processes.
However, it is difficult to imagine how these specific chest postures
could account for more than a small portion of the variability found
in respiratory patterns.

In a thorough investigation of sex differences among normal
subjects Clausen (1951) arrives at the following conclusions: (1) a
general tendency exists for men to have an abdominal type of breathing,
and for women to have a thoracic type; (2) men, on the average, have a
slower breathing rate than women: (3) men, on the average have larger
inspiration - expiration ratio both for thoracic and abdominal breathing
than have women, i.e., women have a greater tendency to show expiratory
pauses and to use a greater fraction of the respiratory cycle for
expiration; and (4) respiration measures are not related to anthropo-
metric indexes, like height, weight, circumference of thorax and abdomen, or Rohren's "fullness of body" index.

In Gray's Anatomy (1948) we are told that: "Costal breathing
predominates in recumbency, and it is said to be more frequent in women,
while diaphragmatic is more frequent in men." (p. 391). The text only
states what has been observed in several studies while the anatomical
basis for these differences is rather underemphasized.

According to Gregor, cited by Halverson (1941), sex differences in respiration, abdominal breathing by boys and costal breathing
by girls, are manifest only after 10 years of age. In contrast to
this, Hagemann, maintains that during the first 12 years of life boys
show a tendency toward thoracic breathing while girls tend to breathe
abdominally. In further contrast to this, Hutchinson, also cited by
Halverson, maintains that costal breathing is predominant in girls of
11 to 14 years of age, while Titz states on the basis of an extensive
investigation of a large number of boys and girls, men and women of
various races, that the respiratory movements of boys and girls differ
very little in character, and that the movements in normal breathing
for both sexes are "fairly equally blanced between chest and abdomen."

Kotikowa and Kotikoff, cited by Haavardsholm (1946), state that no significant sex differences in breathing could be ascertained in the U.S.S.R. Clausen (1951) in his reviews of the literature further maintains that studies of breathing patterns in primitive societies have reported women generally to show abdominal breathing, and that children of both sexes in our culture has been observed to show abdominal breathing. Finally, Weitz, cited by Haavardsholm, maintains that the primary feature in the breathing pattern of Western females is not so much great thoracic excursions as the reduction of abdominal ones.

Summing up the available evidence of sex differences in breathing, we arrive at the conclusion that in our own Western culture a predominance of costal breathing seems to emerge in females at the age of puberty, and that culturally determined postural factors generally seem more important than inherited anatomical ones.

Turning to the next question, what sort of culturally factors are involved, we are confronted with a number of viewpoints.

One hypothesis that has been stated is that the more thoracic breathing frequently found in women in our culture is the result of sex differences in physical exercise, women by and large engaging in less physical exercises than men. The breathing pattern of women in primitive societies has been taken as an argument in favor of this proposition, as well as the observation that sex differences do not seem to occur clearly before the age of puberty. Discussing the hypothesis, Clausen (1951) maintains that it is not supported by his empirical findings. Firstly, no clear cut difference could be assumed to exist with regard to physical exertion in his total sample of males and females, and secondly, no difference was found among male and female subjects with and without systematic physical training.

Another hypothesis suggested (e.g. by Howell, 1930) is that the manner of dressing is the principal contributing cause of sex differences in breathing, women usually wearing tighter dresses than males.

Studying the effect on breathing when women change from tight to comfortable dresses, Hoernicke, cited by Haavardsholm (1946), found that thoracic-lateral movements are quite easily attained, while abdominal movements, if at all attained, will have a tendency to become paradoxical in character. Furthermore, Hoernicke suggests that women breathe more deeply (more abdominally) today than a generation ago.

That tight dresses should be the sole or main cause of costal breathing is doubtful, however. For instance, Hutchinson, previously referred to, in the middle of the last century found costal breathing to be present in girls who had never worn tight dresses. Clausen (1951) too, discusses the hypothesis and finds little support for its validity in his empirical data. Firstly, his female subjects did not wear any tighter dresses than his male subjects generally, and secondly, no differences in amount of abdominal breathing could be found between his male subjects and his psychotic female subjects, the latter group showing more similarity with men's breathing than with his samples of normal and neurotic females.

The third hypothesis to be mentioned is that sex differences in breathing reflects the impact of social norms and ideals. This hypothesis is suggested by Clausen (1951). He states:

"According to the ideal of the feminine figure, the abdomen is not supposed to protrude, hence the costal type of respiration. In order to make the respiration movement as inconspicious as possible they have a brief inspiration followed instantaneously by expiration and this accounts for the low I/E ratio and the sharper peak in the thoracic curve. We would regard the respiration found in males as less influenced by social ideals, and therefore more primitive. The fact that the respiration in psychotic women is more like the men's than like, the two other female groups could be the results of the psychotic women's being out of contact with the social norms and ideals, and consequently having regressed to the more primitive respiratory mode." (p. 63)

Given such a hypothesis, we are forced to ask som additional questions: Does the influence on breathing of social norms and ideals work independent of cultural sex-role typing generally? We know that great differences exist in the socialization of boys and girls in our society, and that the two sexes show differences in a number of behavioral dimensions, in conformity, criminality ratio, body versus field orientation, etc. Do these differences exist independent of the differences found in postural patterns? Clausen' seems to imply that they do so.

This last question brings us over to the fourth hypothesis suggested to account for sex differences in breathing. The hypothesis equates postural factors with affective and motivational ones. A more specific hypothesis within this framework has been formulated by Haavardsholm (1946) as follows:

"The present position of the female in our society is maintained by social norms which exert great inhibitory pressure on women as far as activity and assertiveness are concerned. She is reared more severely than the male and more easily moralistically condemned. Her character. contains more inhibited activity, aggressivity and sexuality... The (supposedly) more abdominal breathing in the modern female as compared with her maternal ancestors may be explained from the point of view of the freer expression of affects and the more expressive activity allowed and obtained by women in the last generation... Sexual sensations derive for a large part from the pelvic area and the orgasm reflex consists partly of contractions of the pelvic and abdominal muscles. That sexual inhibitions will be somatically expressed by a partial immobilization of the abdomen is not unlikely... Thus the question may be asked whether the paradoxical abdominal retraction found in women is due to a muscular contraction whose function is to prevent the natural physiological function of the abdomen during respiration since such a function might give conductance to sexual sensations. No contradictions need essentially exist between the explanation stressing tight dresses as responsible for costal breathing and an explanation emphasizing a culturally determined character structure. If women accept a fashion making it difficult to breathe, it may well be because of characterological features, and when women demand clothes which make it possible to breathe freely, it may well be because women have attained increased self-confidence and greater appetite of the joy of life, and want to feel themselves more free in their bodily movements". (pp. 19-22)

This last hypothesis is very much in tune with Braatöy's and Lowen's train of thought earlier referred to. In ascribing abdominal contractions the role of preventing objectionable impulses from being revived and released Haavardsholm touches upon a viewpoint found in Groddeck's writings. Groddeck (1933) states in his picturesque style:

"In sickness the enmity between the kingdoms of the mind is most clearly revealed. In order to hinder the invasion of the breast and the head by objectionable impulses from the belly, the boundary walls of the diaphragm are first employed, and then the walls and organs in the dome. The patient strives to lock up the forbidden impulses by a temporary, or even by a long-enduring, tension of the musculature of the diaphragm and the space above the navel, as well as blowing out the belly with pockets of air, and in this way tries to prevent them from getting command of the voice of the breast or the wisdom of the head."(pp. 59-60)

Concentrating on Haavardsholm's hypothesis, we are still confronted with a couple of unanswered questions: How can it be explained that psychotic women breathe more like men than neurotic and normal women? How can it be explained that women generally not only breathe more thoracically but also show lower I/E ratio and a sharper peak in their thoracic breathing curve? The former question might of course just as well be answered by reference to a breakdown of sex specific affective defenses as in terms of "being out of contect with social norms and ideals", and the latter question, just as well answered by reference to a tendency to an inconspicuousness in behavior generally as to cautiousness in respiratory movements particularly.

Perhaps the strongest support in favor of a hypothesis explaining differences in respiratory patterns as an expression of postural differences caused by emotional and motivational factors stem from clinical observations and from studies comparing the respiratory pattern of various nosological groups.

Breathing in neurotics, psychotics and normals.

Both Clausen's (1951) and Haavardsholm's (1946) investigations referred to earlier, were undertaken primarily with the aim of throwing light on whether such broad nosological groups as normals, neurotics and psychotics show significant differences in respiratory pattern. Both studies are completely empirical ones, exploratory in nature. In the following we will give a short review of these two studies, both undertaken at the University of Oslo.

Clausen's study is focused on the comparison of respiratory curves - pneumographically recorded - of 50 normals, 40 neurotics and 49 psychotics. His comparisons in terms of means involves a large number of respiratory variables: the rate, the constancy of rate, the inspiration/expiration ratios for thoracic and abdominal respiration, the constancy of these ratios, the size of the thoracic and abdominal amplitudes, the smoothness and regularity of respiration, the occurence of pauses, the shape of the inspiration/expiration transition of the thoracic and abdominal respiration, the shape of thoracic and abdominal inspiration, the shape of thoracic and abdominal expiration, and the relationship between the thoracic and abdominal amplitude and the relative sharpness of their peaks.

In the course of his data analysis Clausen notes a number of differentiating factors: On the average, normals show a somewhat slower respiration rate than neurotics and psychotics; while a fair proportion of normal men have a large thoracic amplitude, in all other groups this is true for only a few or no one at all; compared to normal females, neurotic females more often have extreme thoracic breathing, while psychotic females, more often have abdominal ones; irregular respiration is obtained from some patients, but never from normals; similarities between recordings are more pronounced in normals than in neurotics; while more than half of the normals show s sharper thoracic than abdominal transition, this is practically absent among patients; while a great many patients show triangular abdominal peaks, practically no normals show this characteristic; while some psychotics show dissociative movements of thorax and abdomen, this is absent in normals and neurotics.

A most interesting aspect of Clausen's study is his attempt to formulate a sign scale by which normals and neurotics may be operationally differentiated. See next page. Altogether he arrives at 12 differentiating signs for males and 10 for females. By giving the signs with a seemingly marked differentiating power a weight of 2, and the rest a weight of 1, Clausen arrives at a male scale with a potential range of scores from 0 to 20 points, and a female scale with a range from 0 to 16 points. Rescoring his data according to these scales he finds that 75% of neurotic males and 70% of neurotic females fall outside the range of variation for their normal control group. This is indeed a remarkable and encouraging finding.

Clausen's Sign Scale of Respiratory Items.

MALES	FEMALES
1. Respiration rate higher than 14.5 cycles per min. (2 pts)	1. Respiration rate higher than 15.8 cycles per min. (2 pts)
2. Respiration curve irregular (1 point)	2. Respiration curve irregular (1 point)
3. Dissimilarity between recordings (2 points)	3. Dissimilarity between recordings (2 points)
4. Size of thoracic amplitude small (1 point)	4. Thoracic inspiration notably convex (1 point)
5. Thoracic inspiration notable convex (1 point)	5. Abdominal inspiration rectilinear (2 points)
6. Thoracic peaks bell-shaped (2 points)	6. Abdominal peaks triangle-shaped (2 points)
7. Thoracic inspiration-expiration transition neither sharp nor blunt (1 point)	7. Abdominal inspiration-expiration transition sharp (2 points)
8. Abdominal amplitude small (1 pt) if very small (2 pts)	8. Abdominal peaks sharper than thoracic peaks (2 pts)
9. Abdominal inspiration rectilinear (2 points)	9. Abdominal expiration slightly concave (2 pts)
10. Abdominal peaks triangle-shaped (2 points)	10. Abdominal amplitudes greater than thoracic amplitudes (1 point)
11. Abdominal inspiration-expiration transition sharp (2 points)	
12. Abdominal peaks sharper than thoracic peaks (2 points)	

Clausen is fully aware of that his signs are in need of an independent cross-validation. He attempts two indirect approaches to this question. First, by comparing psychotics with his normal and neurotic samples. Since no initial effort was made to construct the scale as to differentiate between normals and psychotics, he finds it of considerable interest hat his psychotic group show averages close to those found in his neurotic groups. Next, Clausen picks out from among his neurotic subjects two extreme groups on the sign scale and compares their clinical pictures. In his group of low

scorers, he finds a distribution of 3, 6, and 1, rated as mildly,
moderately and severely neurotic respectively, while in his group of
14 high scorers, a total of 7 were rated as severely neurotic.

As stated, Clausen is fully aware of that these indirect
approaches are insufficient to validate his respiratory signs. The
approach, however, attracts interest and encourages further investigations.

One may ask, how can the hypothetical differentiating power
of the various signs be explained. Do they fit into an overall theory
of neurotic mechanisms? Clausen points out that his signs are far
from completely independent of each other. The sign: "Thoracic peaks
bell-shaped" overlaps to a certain extent "Thoracic inspiration notably
convex" and "Thoracic inspiration-expiration transition neither sharp
nor blunt." The same is true for the sign: "Abdominal peaks triangle-
shaped," which overlaps "Abdominal inspiration rectilinear," and
"Abdominal inspiration-expiration transition sharp." In fact, a single
factor, the relation between inspiration-expiration transition for the
thoracic and abdominal peaks, synthesizes a number of the differences
found; the transition for abdominal peaks by and large being considerably sharper among neurotics than among normals.

In commenting on his results, Clausen pays special attention
to this last feature:

"Characteristic of respiration in emotionally disturbed patients
is a shorter period for maximal contraction of the diaphragm than
is found in normals. This inability or unwillingness to keep the
diaphragm contracted long enough to make the transition from inspiration to expiration a gradual one, might be a result of the
condition of muscles in the abdominal wall. When the diaphragm
is contracted by inspiration the abdominal muscles have to be
relaxed to compensate for the increased pressure in the abdomen.
The rapid transition may be due to inability or unwillingness to
relax reciprocally the muscles in the abdominal wall. It is
possible that the diaphragm of the neurotics work against an
increased abdominal pressure, thus decreasing the duration of
maximal diaphragmatic contraction. Whether the disturbance would
be primarily in the diaphragm or in the abdominal musculature
we have no way of telling from this study." (p. 62)

Clausen provides no explanation for the greater inconsistency and the higher inspiration ratio found in neurotics, neither
does he deal at any length with the differences found among neurotic

and normal males in amplitude and in thoracic respiration, the latter
factor being more bell-shaped, probably more restrictive and cautious
in the former group. In general, Clausen's conslusions are very
circumstantial and he points out time and again that his diagnostic
indices are unsuitable for the evaluation of individual subjects.
This, of course, is self-evident taking into account that nosological
groups such as normals, neurotics and psychotics, are highly hetero-
geneous groups.

- - - -

Haavardsholm's investigation focuses heavily on a comparison
of EMG potentials from above the xiphoid process in four small groups
of subjects, normals, neurotics (inpatients in a psychiatric hospital),
psychotics (also inpatients) and healthy (subjects considered positively
healthy by their former therapist). His groups are all very small, from
3 to 11 subjects with an average of 7 subjects, and his results
extremely tentative. The variables compared consisted of the average
EMG inspiratory amplitude, the average EMG expiratory amplitude, the
inspiratory-expiratory amplitude ratio, and the respiratory rate. Each
subject was tested two or three times, and a substantial test-retest
consistency was found in nearly all cases.

Haavardsholm's findings give raise to the following suggestions:

1. The more mentally ill a person is a) the less is the difference
   between his inspiratory and expiratory action potentials, b) the
   higher is his action potentials during expirations, and c) the
   higher is his respiratory rate.

2. In terms of discriminatory power, the respiratory rate seems to
   represent the less promising variable, and the inspiratory-
   expiratory EMG ratio the most potent one.

3. When a not severely mentally ill person shows little difference
   between inspiratory and expiratory action potentials, his action
   potentials are fairly low generally and consistently lower than
   in severely ill persons.

4. The pattern of EMG potentials found in severely ill persons
(schizophrenics) can be approximately duplicated in normal subjects
by the administration of pain stimuli together with an instruction
to suppress all expression of pain.

5. Among severely ill persons (schizophrenics) the inspiratory circumference changes of the body are smaller than in normals with
inspiratory action potentials of approximately the same size, the
reason probably being that the diaphragm and the abdominal wall
in the former group operate as antagonistic muscles both being
contracted permanently or periodically at the same time. This
point corresponds, of course, to some of Clausen's findings
previously mentioned.

6. Neither a smooth nor a concave expiratory pneumographic curve
necessarily implies a passive expiration, although an irregular
and a convex or rectilinear curve implies an active contraction.
Accordingly, EMG recordings represent a far more reliable method
to get at a subject's ability for expiratory relaxation than
pneumographic tracings.

7. A completely passive expiration does probably not exist. The lack
of action potentials sometimes found during expiration refers to
its latter part only. The question is not if an expiration is
accompanied by action potentials or not, but to what extent it
moves over from an active to a passive phase before the next
respiratory cycle comes into play. This observation has been
substantiated by later more systematic studies by other investigators.

8. Physical exercise and exhaustion have variable effect on respirations. In some persons it produces a more regular respiratory
curve and greater expiratory relaxation, in others not. This
finding may of course have some bearing on the large individual
differences found in attitudes toward athletics and physical
exercise.

As stated, Haavardsholm's study suffers from a number of
shortcomings. His samples of subjects are small, his subjects are
not matched on a number of significant variables, and so on. The

study has to be considered mainly as a preliminary investigation within the framework of an hypothesis formulating endeavor. This is also
very much the way Haavardsholm considers his empirical material. His
data presentation is partly focused on case histories with discussions
of the relationship present between EMG potentials and individual
personality characteristics. Of interest is the potential sign scale
implicitly referred to under points 1, 2 and 5 above. In this respect
Haavardsholm's investigation shows some of the same features as
Clausen's study, although it is much more modest in its scope and
intentions. The two studies considered together, complements each
other and points towards an area of research which so far has been
rather underdeveloped, but which, in view of these studies, offer great
promise for further exploration.

Breathing in schizophrenia and in schizoid characters.

Corwin and Barry (1940) in a study comparing respiratory
tracings of schizophrenics and normals found that expiratory plateaux
of 5 seconds or more occur much more frequently among schizophrenics
than among normals. In one schizophrenic patient they found plateaux
with durations up to 18 seconds. In fact, plateaux of 10 seconds and
more appeared only in the records of schizophrenic patients.

The same observation has been reported by Paterson (1934)
who states that recurring breath holding seems to be characteristic
of some schizophrenic patients, and also by Alexander and Saul (1940)
who note that "sudden breath holding of over 15 seconds has appeared
only in the records of psychotics and very severe neurotics."

The respiratory plateaux often found among schizophrenics
don't necessarily imply irregular breathing. In fact, several studies
(Thompson and Corwin, 1942; Paterson, 1934; Wittkower, 1934; Thompson,
Corwin and Aste-Salazar, 1937) have all pointed out that regular
breathing seems to be more common among schizophrenics than among
normals. Although Finesinger's (1943) results question whether
schizophrenics show higher regularity than normals on all facets of
breathing, he finds his schizophrenic group to show a lower mean over-
all irregularity score than anxiety neurotics, other neurotics and
normals, and at least on one dimension, statistically significant

higher regularity than among normals.

Neither does the observations of breath holding tendencies in schizophrenics imply a general tendency to slow breathing. Again, a number of studies (Paterson, 1934; Wittkower, 1934) have emphasized that schizophrenics often have a higher respiration rate than normals. For instance Wittkower (1934) states:

"Comparison of the findings ascertained in normal subjects and schizophrenics shows unequivocally that the schizophrenics breathe more frequently...rates of more than 22 respirations per minute observed in schizophrenics cannot be noted in the normal subjects. Particularly distinct,...is the difference when considering the women...15% of the female normal subjects and 41.25% of the female schizophrenics breathe more frequently than 18 times per minute." (p. 699)

One may certainly ask how can a tendency to respiratory plateaux on the one hand be compatible with a tendency to rapid breathing on the other, and furthermore, how can these seemingly inconsistent tendencies be compatible with a tendency toward regular breathing?

In order to throw some light on this question we will have to remind ourselves that schizophrenia is a rather ill-defined nosological entity, that it includes different subsyndromes, and that it is highly questionable whether each of these subsyndromes refer to one and the same underlying disease process.

Paterson (1935) emphasizes that a characteristic feature of schizophrenics is an "extreme regularity of rhythm broken by a temporary extreme irregularity, followed again by a very regular rhythm." As regards the type of irregularity periodically found he mentions breath holding and a curious tremor of the respiratory muscles. And he gives the following suggestion as to the meaning of these periodical irregularities:

"It not infrequently occured in one of these experiments that while a patient was breathing regularly he suddenly started to breathe irregularly. When questioned, he would say that he heard a voice speaking to him, even although his lips had not been observed to move. On more than one occasion this occurred where the patient had denied hallucinations, but later admitted them, when asked about his thoughts at the moment when his breathing changed. On other occasions sudden irregularities of breathing were found to be accompanied by olfactory hallucinations." (p. 49)

Paterson's viewpoint implies that we should expect the more
irregularity in breathing in schizophrenic patients the more these
patients are in a state of active hallucinations, and perhaps also,
in a state of fright and terror by primary process thinking intruding
into their consciousness. On the other hand, the more the schizophrenic patients were demented, the more regular we would expect their
breathing to be.

Paterson (1935) presents empirical data substantiating such
a hypothesis. Among schizophrenics hospitalized less than one year he
found 38% only, showing an extreme regular type of breathing, while
when considering schizophrenics being hospitalized for more than 5
years, the percentage increased to 82%. Summarizing his findings, he
states:

> "Examination of the schizophrenic group showed that the irregular
> breathers were mostly to be found among those who had been ill
> for less than five years...the percentage of regular breathers
> goes up according to the length of time that the patient has been
> in the hospital...where there is considerable dementia...the res-
> piration is usually regular." (p. 45)

As regards breathing rate, it has been noted by several
psychiatrists that patients in catatonic states show slower rate than
other schizophrenics. For instance, Paterson (1934) states: "With
regards to patients in a more severe catatonic state, Gullotta has
remarked that catatonics breathe more slowly than other schizophrenics.
My observations support their view." Kempf (1930) elaborates on the
same point as follows:

> "The catatonic manner of respiration seems usually to be
> characterized by a shallow, abdominal amplitude and decreased
> rate. It has an insidious stupor to sleep producing effect with
> probably a slowing up of endocrine functions, as the affectivity
> retreats from reality. We should contrast it to the increased
> amplitude and rate of the thoracicodiaphragmatic breathing and
> contacted abdominal muscles in man and the higher animals when
> exited and aroused by danger... In catatonic we have shallow,
> abdominal amplitude on relatively deflated lungs, with almost
> imperceptible thoracic amplitude, with probably a minimum of
> oxygen in the blood." (p. 183)

Alexander and Saul (1940) points to somewhat similar
features in hebephrenic patiens: shallow breathing, low chest level
and absence of rounded tips in the respiratory tracings. Other ob-

servers too have noted shallow breathing as a characteristic feature
of schizophrenic patients. For instance, Paterson (1934) concludes
that "it appears that schizophrenics breathe more shallowly... than
normals," and Wittkower (1934), that "often the respiration of the
schizophrenic is abnormally shallow..." In an empirical study by
Thompson and Corwin (1942) it is again pointed out that a "smaller
volume of tidal air" is characteristic of schizophrenics. In fact,
the latter authors claim that they have been able to discriminate
extremely well between schizophrenics and normals on the basis of the
following two criteria: greater regularity and smaller volume of
tidal air. Also Lowen (1958) emphasizes low air intake and shallow
breathing as characteristic of schizophrenics and prepsychotic
schizophrenic characters, although his description of the typical
respiratory pattern deviates sharply from the one suggested by Kempf.
Lowen states:

> "...the respiration of the schizoid character and of the schizo-
> phrenic shows a characteristic disturbance...low air intake in
> spite of...soft chest and seemingly large excursion of the rib
> cage. There is another factor involved in this paradox. In
> the schizoid structure the expansion of the chest cavity is ac-
> companied by a contraction of the abdominal cavity. This prevents
> the diaphragm from descending...Under such a condition, the
> schizoid or schizophrenic makes an effort to breathe in the upper
> part of his chest in order to get sufficient air...the diaphragm
> is relatively immobile; it is frozen in a contracted condition...
> Since the diaphragm is inactive, a strong expansion of the chest
> cavity tends to pull the diaphragm upward by suction... One ob-
> serves that the belly is sucked in during inspiration, then pushed
> out during expiration. This is not the normal type of respiration.
> In the average individual chest and belly tend to make the same
> movement." (p. 337)

As regards Lowen's observation of schizophrenics' paradoxical
respiratory movements, i.e., that the belly is pulled in during inspiration, then pushed out during expiration, we want to quote a para-
graph from Clausen's (1951) monograph:

> "We have found (in normals) that the thoracic and the abdominal
> movements are running parallel. Inspection of the respiration
> curves seem to indicate that the movements are synchronous, so
> that the inspiration and expiration phases start at the same time
> in thorax and abdomen. In psychiatric (psychotic) cases we have
> seen instances where the two sets of movements are dissociated,
> but this is a very rare occurence and dissociative movements most
> certainly are not a characteristic feature for respiration of
> psychotics." (p. 53)

Clausen doesn't give any information regarding the type of psychotics he found to show paradoxical breathing movements. He informs us that his psychotic groups are not exclusively schizophrenic, but dominated by schizophrenic patients. This being the case the "very rare occurence" of paradoxical breathing found, weakens the general validity of Lowen's hypothesis. So does also Clausen's finding that "psychotic females more often than normal females show an abdominal type of breathing." However, although a majority of Clausen's psychotics were schizophrenic patients, a large part evidently were severely demented, while Lowen seems to focus much more on incipient schizophrenics, on schizophrenics in active struggle, panicky and fright. According to our earlier suggestion we would expect to find significant differences between these two states. In fact, Clausen alludes to the same point of view. When commenting upon his empirical data, he states:

"The sharper inspiration-expiration transition in abdominal respiration is not characteristic of mental disturbance per se, in that it does not occur regularly in psychotics. However, as we found in the psychotic groups some patients with symptoms of emotional stress, whereas others were completely indifferent and apathetic, it seems that this feature may be related to the expression of emotional conflicts...patients with emotional conflicts have a modified respiration..." (p. 62)

What Clausen alludes to here seems to be that only in the case of psychotics with symptoms of emotional stress do we find a modified respiration in terms of a sharper inspiration-expiration transition in abdominal respiration, i.e., a feature to be expected if the expansion of the diaphragm or the chest cavity is accompanied by a contraction of the abdominal wall. We are reminded of Alexander and Saul's observation of absence of rounded tips as a characteristic feature of hebephrenic's spirograms, and of Haavardsholm's observation of disproportional high EMG potentials from above the xiphoid process, as related to the abdominal circumference changes actually taking place in psychotic subjects. And we are lead to go one step further in our hypothesizing: that as the tensional state of the diaphragm and the abdominal muscles increase, a point will be reached where abdominal respiratory movements will disappear and eventually turn over into a dissociated movement pattern, that is, become opposite

and dyssynchronized with thoracic respiratory movements. This
phenomenon we may assume would correspond to a florid and flamboyant
schizophrenic symptomatology, accompanied by relatively irregular
breathing and a relatively fast breathing rate. As deterioration of
psychic processes and affect detachment increases and as a state of
affect poverty and dementia emerges, we might expect parallel respiratory changes to take place, so that in the completely affect detached
schizophrenic, the catatonic for instance, we end up with a respiratory
pattern characterized by abdominal predominance, slow and extremely
regular breathing. Clausen's findings, combined with Kempf's,
Patterson's and others observations, strongly suggest such a
variability within the schizophrenic pattern, a variability followed,
however, by a consistent relatively low intake of air.

The hypothesis just stated, has as far as we know, never been
systematically tested. It implies in a sense, not a distinctive
schizophrenic breathing pattern, but a variation corresponding to
variation in affectivity or activation going from one extreme to
another.

- - - - -

One may ask if the same variation is present in normal subjects, in normals varying along a dimension from an introvert (schizoid)
to an extrovert (cyclothymic) style of life. Thompson and Corwin (1942)
have presented some data related to this question. In one study they
found that Southern Negroes, by and large show smaller volume of
tidal air and more regular breathing than Whites. In attempting to
explain these differences they emphasize the Southern Negroes' "general passivity, intense imaginative tendencies, amounting often to frank
hallucinations, religious manias and extensive suspicion and superstitions." These traits possibly being a concomitant aspect of their
introverted breathing pattern. In another study, Thompson and Corwin
again focus on a group of normal subjects, this time to see how far
it is possible to predict a personality disposition on the basis of
respiratory data alone, that is, on the basis of a respiratory index
taking into account regularity and volume of tidal air the smaller
the volume and the greater the regularity, the greater the likelihood
of a schizoid disposition. In order to arrive at relatively reliable

personality diagnosis, each subject was evaluated on the basis on an average of about eight hours of psychiatric interviewing. In both the psychiatric evaluation and in the scoring of the respiratory records each subject was placed in one of the following diagnostic categories: introvert, extrovert, miscellaneous, and neurotic, the last category being defined respiration-wise by marked irregularity with no discernible pattern. The percentages of agreement found for the various categories were 80, 75, 88, and 82, respectively. The number of subjects in the various categories were 20, 16, 13, and 6, respectively. Of course, this is fairly small groups. However, the consistency of the findings indicate a fairly high level of significance and fully justifies Thompson and Corwin's conclusion that "the mode of breathing is an index to some of the personality components of normal persons."

Breathing in anxiety states, in hysterical and compulsive neurotics.

Approaching the problem of the prevalent breathing patterns in anxiety neuroses, and in well-compensated neurotic character structures, from the point of view discussed in the last section, we should not only expect marked differences, but differences going in a specific direction -- patients in the former group showing more thoracic, more irregular and more arrhythmic breathing than patients in the latter group.

We just mentioned Thompson and Corwin's definition of neurotic breathing (referring to anxiety states and phobias) as consisting of marked irregularity with no discernible patterns. Perhaps the most systematic study made so far of respiratory irregularities in different nosological groups, is Finesinger's (1943) investigation concentrating on 7 different aspects of irregularities. Of the different groups studied, anxiety neurotics, other neurotics, schizophrenics, and normals, the former group was found to show greatest irregularities on all variables; on six of the seven variables were statistically significant differences found with schizophrenics, and on four of the six variables, significant differences with normals.[1]

In describing a case history of a 22-year-old woman suffering from a lasting anxiety state, with daily heart sensations and apprehensions, Braatöy (1947) states:

"In a supine position, she draws her abdomen inward and pulls her breast upwards. Asked to 'breathe with the belly' she cannot. She writhes her body awkwardly, pulls her breast still higher, while her abdomen is kept back as previously. She understands that it is something peculiar with this phenomena but cannot explain what is wrong. Asked to tense her belly and make 'big belly' as children do, she cannot. Her abdomen is still kept back. Asked to push or 'bear down' with an effort similar to that of defecation, her abdomen does still not protrude. Without any instruction, she is lying tense with short, superficial, shallow, high-costal breathing." (p. 193)

An anxiety neuroses, the presence of manifest anxiety, either diffuse or phobic in nature, will always, according to Braatöy, go along with typical breathing difficulties (clearest seen with the patient in a supine position): abdomen kept back, chest arched and drawn up, the breathing being short, jerking, and shallow. He assumes very much the same pattern to be characteristic of hysterical patients: "The hysterical (patient)....is in active conscious and preconscious tension. The tension is manifested in...arrhythmic breathing frequently high costal."

Also Christie (1935) maintains that anxiety neuroses is characterized by specific features. Christie differentiates between various forms of irregularities; irregularities 1) of respiratory depth, 2) of respiratory level, and 3) of respiratory rate. Typical of anxiety neuroses is a tendency to rapid and shallow breathing, to irregularity of respiratory level and respiratory dept, and, although to much lesser extent, to irregularity of respiratory rate. Christie further maintains that conversion hysteria show many similarities to anxiety neuroses, but not a completely identical breathing pattern. Sighing respiration may occur in anxiety neuroses, but not in the form of a series of deep sighing respirations, the latter being viewed as

---

1) Finesinger further reports that the incidence of sighing respiration seems to differentiate quite well between anxiety neuroses and normal controls, the incidence for a 6 minute period being 60 and 21% respectively, and he concludes that "it seems safe to predict the presence of a neuroses in a person whose tracings averages more than one sigh per minute."

more less specific for conversion hysterias. However, the main
differentiating characteristic is considered to be the depth of
breathing. Although the breathing pattern of conversion hysterias
might be just as irregular as in anxiety neuroses, it would always
be deeper and less shallow, Christie maintains.

Very much in contrast to the respiratory patterns mentioned
so far, the compulsory neurotic is described by Braatöy (1947) as
follows:

> "A compulsory neurotic's breathing pattern is a self-imposed,
> lawful pattern. It must not be changed involuntarily. Con-
> sequently, anger, eagerness and affect involvement have to be
> avoided. In affect the possibility exists that the breathing
> will be carried away so that it cannot any longer be controlled
> or directed... Hence, the compulsory neurotic is vigilant in
> relation to affects, and his breathing is a prepared-restrictive
> one of quite another type than the one characterizing the anxiety
> neurotic. The latter is immediately retentive, because the
> anxiety tenseness draws the breathing in, tightens the neck,
> shoulder and arm muscles, and thereby fixates the breathing in
> a high-costal pattern. The breathing of the compulsory neurotic
> is "mediate," voluntary retentive, and may as such have different
> forms. Frequently, it will be modulated by a specific back-
> posture, pressed down in the abdomen, because such a person has
> learned that breathing can best be controlled by the diaphragm
> and the abdominal muscles." (pp. 217-218)

In short, Braatöy maintains that the anxiety neurotic breathes
fast and irregular of anxiety, while the compulsive neurotic breathes
in a certain way, usually with an abdominal predominance, in order to
avoid anxiety. Many compulsory neurotics make use of various forms
of conscious breathing ceremonials. Whether this is the case or not,
the compulsory acts will always have elements of something cermonially
fixated. In Braatöy's (1947) words:

> "Affects have to be avoided because one may be carried away by
> them and thereby forget ones ceremonial commitments. Thus, among
> all compulsory neurotics one will find the retentive, reserved,
> controlled mode of breathing." (p. 218)

To illustrate this point we may quote a section from Braatöy's
(1947) case history of an overcontrolled and inhibited young physician:

> "His respiration was not costal with arched chest and short inspira-
> tion and pulled-in abdomen. His chest was flattened with small
> respiratory movements, while he was breathing much more clearly
> with his abdomen. He, himself had the feeling that his breathing
> was not free. The movement of the chest was strikingly small.

He breathed with a flat chest as if a weight had been placed upon it, and when the respiratory movements increased during relaxation-- or if asked to breathe deeply--a hissing sound could be heard from nose and lungs... In this case, abdominal movement increased in amplitude while the chest followed sluggishly and flat in the respiratory movements. One gets the impression that his respiration--specifically under inspiration--is met by a tough, sucking resistance." (p. 209)

Braatöy's assertion that a compulsive controlled character structure goes along with a retentive, reserved and controlled mode of breathing fits well observations by Alexander and Saul (1940), observations leading them to postulate that there exist a close relationship between the main mode of functioning on the psychological and physiological level.

In short, Alexander and Saul maintain that "psychological tendencies are likely to influence those physiological functions which have the same vector quality." By vector quality they refer to the direction, in relation to the individal, of tendencies which may be observed both psychologically and physiologically. They differentiate between three vector qualities in all, between 1) centripetal (incorporating), 2) centrifugal (eliminating), and 3) retentive qualities. If one of these vector qualities is predominant of the emotional or psychological level, it will also be predominant in the respiratory (as well as in the gastrointestinal) area. According to this hypothesis we should expect a receptive character to show inspiratory "spikes", i.e., (sudden deep inspirations breaking the usual rhythm), and expiratory "hooks," i.e., slight hesitation in expiration; an expulsive character to show litte supplementary air, i.e., deep expiration, rounding of expiratory tips, and possibly a tendency to expiratory plateaux; and a retentive character to show slight hesitation in expiration, lack of deep breathing, no rounding of expiration tips and possibly a tendency to a controlled and reserved mode of breathing.

Alexander and Saul's hypotheses represent a form of a psychosomatic interactional theory. In spite of its wide-ranging implications, it is a fairly restricted theory. Its three vector qualities remind us if Erikson's (1950) conceptualization of psychosexual developmental modes. However, Erikson's model is more inclusive. He differentiates not only between two forms of incorporative modes, but also adds a fifth mode, the intrusive one. Although the operational definitions stated by Alexander and Saul are rather vague and discursive, their approach as a whole certainly represents a courageous attempt to formulate a general explanatory principle for variations in respiratory patterns.

So far as we know, no systematic attempts have been made to test and evaluate empirically Alexander and Saul's respiratory hypothesis.

- - - - -

Turning back to the breathing pattern of obsessive-compulsive personalities, to Braatöy's conception that anxiety neurotics breathe of anxiety, while compulsory neurotics breathe in order to avoid anxiety and excitement, one would have to ask how can this latter mechanism be brought about? The mechanism in question, according to Braatöy, consists of a general tensing of skeletal muscles. While the hysterical patient and the patient related to hysteria show specific resistances, specific muscle pains and other local spasm, so in the compulsive neurotics one meets a general defense and a fairly general non-localized rigidity. The compulsive character is a person with a heavy armor both on the psychological as well as on the somaticomuscular level. The essential aspect of this armoring is that it binds affects and anxiety, that it protects the person from both external and internal dangers and threats.

We find very much the same train of reasoning in Lowen's (1958) theoretical elaborations. Lowen too emphasizes that muscular tension in itself is not necessarily an armoring. For instance, in both schizophrenics, oral and masochistic characters we find severe muscle tensions, without these tensions constituting an armor. The essential criterion of an armoring, an efficient binding of anxiety, is, according to Lowen, that "the front of the body is hard." It is the hypertonicity of the chest wall in particular that is essential to an armor.

In a case history of a young man in his late twenties with a typical oral character, Lowen has the following to say about his breathing:

"Respiratory movements were limited to the chest, which seemed
mobile. However, the shoulders did not participate in these
movements. Neither the inspiration nor the expiration involved
the abdomen to any visible extent, the diaphragm retained its
contracted position." (p. 149)

Braatöy (1947) too has commented upon the breathing patterns
of oral characters: "it is a near connection between oral characters
relation to pleasure in life and their way of breathing, in so far as
one has to breathe in a certain way when one is sucking." Specifically,
Braatöy suggests "small, relatively short amplitudes, and an uncertain,
semi-irregular lack of sure rhythm" as typical for oral characters.

In describing the physical apperance of a young man with a
masochistic character, Lowen states: "One noted that the chest moved
easily with respiration and the abdomen was soft." But also that
"when the patient attempted to make a deep expiration, as in a deep
sigh, the chest relaxed but the descending wave piled up into a hard
knot in the middle of the abdomen." (p. 178). Braatöy (1954) too in
a case description of a masochistic character notes the mobility of
the chest, but also a sudden periodical interruption of every deep inspiration by the drawing in of the abdominal wall -- a pattern probably
related to sucking inspiration.

In describing a phallic-narcissistic, semi-compulsive character, "a young man in his early thirties, attorney by profession, fairly
successful in his work and confident in his future," Lowen states:

"He was self-confident, outspoken and on guard... He told me
that he always engaged in sports and had continued his athletic
activity into the present. His face was expressive, the eyes
lively and open, the mouth easy and tending to smile, the jaw
rather set and determined. On the couch, one noted the wide
shoulders, full chest, narrow hips and rather tight legs. Respiration tended to be abdominal, the chest was held in the inspiratory
position." (p. 261)

In describing a man about forty years old with a passive-
feminine character, Lowen states:

"He had a soft, modulated voice and a gentle manner. There were
no hard lines on his face...he was very polite and cooperative...
his chest was tight and did not move with respiration. Since
the abdominal muscles were also tight, respiration was greatly
reduced... The tensions were located deep in the intercostal
muscles..." (p. 282)

The two last patients described, although different in many
respects, both exemplify according to Lowen, armored character stuctures,
while the two former ones, also different from each other in many
respects, exemplify non-armored structures. The important and crucial
difference between armored and non-armored neurotic personality
structures is the developmental level at which the predominant dynamic
nucleus of their character organization was established, Lowen maintains.
The formation of an armored character presupposes that the major personality conflicts emerged after the person as a child had reached a
post-oral developmental level, that is, after he had attained a firm
anchorage to an erect posture, that is, usually not before the age
of two or three.

Lowen offers the following speculations as to the dynamics
behind the formation of a psychosomatic armoring:

> "One can hypothesize that man's erect posture, exposing this
> (front) aspect of the body, required the development of an
> armor for its protection.. This might appear to be the case
> except that the freeing of the fore limbs for aggression in
> the human animal, as compared with their use for support and
> locomotion in the lower mammals, balances the scales. As long
> as the arms are available for attack and defense, there is no
> need for armor. Genetically, the armor develops through the
> immobilization of the aggression in the child. Psychologically,
> the armor is the expression of the attitude of stiffening to
> meet an attack rather than striking back. Dynamically, the
> tension in the front is produced by pulling back the shoulders
> and pelvis, thus putting all the front muscles on the stretch
> at the same time that they are contracted. When the front and
> back of the body are thus encased in a rigid sheath of tight
> muscles,we can say that the organism is armored." (p. 232)

The distinction between a compulsive character and other
armored character is one of degree only, Lowen maintains: "...a
compulsive behavior is the extreme form of rigidity on the psychological level. The same excessive rigidity characterizes the somatic
structure...it is (merely) more severe and penetrates more deeply
toward the center."

How does the armoring bind anxiety? "It does so,"Lowen
states, "by reducing respirations through an unconscious control over
the muscles of the front of the body. Although the diaphragm is
relatively free, the rigidity of the total structure (the tense chest
wall, etc.) limits the intake and output of air."

This last viewpoint, that anxiety becomes bound by a restriction of intake and output of air, i.e., through inhibited respiration, is in complete contradiction to Perls et al. (1951) anxiety hypothesis. According to these authors, anxiety is a symptom which develops in all situations where the organism blocks its breathing and is deprived of adequate oxygen:

"Anxiety is the experience of breathing difficulty during any blocked excitement. It is the experience of trying to get more air into lungs immobilized by muscular constrictions of the thoracic cage." (p. 128)

Consequently, far from viewing low air intake as a binding of anxiety, the phenomenon is considered as a central factor in the production of anxiety.

It is interesting to note that the very same contradiction is apparent in Reich's (1942) writing on the same subject. One place he states:

"The inhibition of respiration, as it is found regularly in neurotics, has biologically speaking, the function of reducing the production of energy in the organism, and thus, of reducing the production of anxiety." (p. 276)

Later, twenty-five pages later, in the same chapter in the same book, he writes:

"The inhibition of respiration - specifically, of deep expiration - creates a conflict: the inhibition serves the purpose of damping the pleasurable excitation..., but just in doing so, it creates an increased susceptibility for anxiety..." (p. 301)

To answer which one, if any, of the two conceptions is the most valid one, we have to turn to the results of empirical investigations.

- - - - -

Studying the respiratory efficiency of a sample of mental patients suffering from anxiety and with anxiety as the predominant symptom and a normal control group, Mezey and Coppen (1961) found that the respiratory effort required to extract a certain amount of oxygen from respired air was significantly higher in the former than in the latter group. They state:

"The results show that under resting conditions, whether lying or standing, the respiratory efficiency of anxious patients is impaired... Since the respiratory abnormalities were reduced after clinical improvement it is unlikely that they can be attributed to long-standing or constitutional physiological peculiarities. There is the possibility that these abnormalities may be related to an increased activity of voluntary mescles."
(p. 174)

In another investigation by the same authors (1960), it is pointed out that sodium amytal in non-hypnotic doses causes a temporary reversal of the respiratory abnormalities observed in anxious mental patients, and furthermore, that their respiratory function under the influence of sodium amytal becomes similar to that found in normal subjects, and in the very same patients themselves after clinical recovery.

Mezey and Coppen suggest that the respiratory inefficiency observed in anxious states is caused by shallow breathing, that is to say, they suggest that inhibited shallow respiration is not reducing anxiety but quite the opposite, producing it.

Although Mezey and Coppen do not elaborate on their viewpoint, we would like to ask what sort of mechanism is probably involved in the respiratory inefficiency of anxiety states? In our earlier discussion of the respiratory apparatus we touched upon the observation that in deep breathing in normal subjects about three-quarters of the ventilation is due to diaphragmatic movements, that is to say, a blocking of diaphragmatic excursions have much more grave consequences on ventilation than a blocking of the mobility of the thoracic wall. This also follows from the fact that deep diaphragmatic breathing is much more efficient in terms of oxygen extraction than high costal breathing because a proportionally greater part of the air inspired reaches the pulmonary alveoli of the lungs. These considerations lead up to the inference that anxiety most probably is caused by shallow breathing in the sense of high costal, non-diaphragmatical breathing. But it also leads up to another inference, namely that by keeping the chest wall in an armored, fixated position an individual will force himself, so to speak, into a diaphragmatic type of respiration, while if the chest is mobile, a partial diaphragmatic spasm not only may occur but also gradually force respiration into a comparatively inefficient high costal stratum.

Substantiating such a viewpoint are some very interesting speculations by Perl et al. (1951). They state:

"The neurotic who suffers anxiety-states simply cannot...breathe - for, unaware of what he is doing and therefore uncontrollably, he maintains against his breathing a system of motor tensions, such as tightening the diaphragm against tendencies to sob or express disgust, tightening the throat against tendencies to shriek, sticking out the chest to appear substantial, holding back the aggression of the shoulders, and a number of other things... He is utterly incapable of a complete, unforced exhalation. Instead, his breath comes out in uneven spurts -- staircase breathing -- and it may stop, as if bumping into a wall, long before thorough emtying the lungs. After hitting such a wall, he may then, by forced contraction, expel more air, but this is artifical and continues only as long as the deliberate effort." (p. 131)

And discussing how to get relief from acute anxiety, they maintain: "...a partial relief of any given instance of anxiety can be obtained, paradoxically, by the tightening even further the narrowness of the chest instead of resisting it." By this narrowing of the chest the person forces himself to change from a thoracic to a more abdominal (deeper) type of breathing. Of course, this measure does not result in any permanent changes in anxiety proneness. In order for that to occur a modification of the underlying muscular hypertonicity has to take place, and such a modification would imply a release of the excitement and affects being arrested by these very hypertonicity patterns.

In concluding, the preceding discussion leads us to reject Lowen's idea that low air intake and output blocks anxiety, although on a hypothetical basis, we tend to agree with him that a chest armoring may subdue and prevent anxiety by chronically reinforcing diaphragmatic excursions. In the same way, we tend to agree with Reich's viewpoint that inhibition of deep respiration, i.e., diaphragmatic respiration, creates an increased susceptibility for anxiety, while, at the same time, rejecting his hypothesis that respiratory inhibitions by means of reducing energy production, reduces anxiety. As stated, we tend to agree that inhibition of thoracic respiration may reduce anxiety proneness, however, not by the reduction of air intake but by reinforcing diaphragmatic breathing. In short, our hypothetical position coincides intimately with Braatöy's opinion,

that an anxious person breathes of anxiety and a well-compensated person in such a way as to avoid anxiety.

The preceding discussion re-emphasizes some of the viewpoints
launched in earlier sections. In discussing the breathing patterns
of schizophrenics we suggested a dimension going from high costal to
deep diaphragmatic breathing, a dimension corresponding to a high
versus low level of activation and flamboyant symptomatology. In
this last section we have followed through with the same train of
reasoning in the field of neurotic disorders, and our review of the
literature has brought out a somewhat similar dimension in this area.
In the very last part of this section we have attempted to describe
some of the postural mechanisms which may be responsible for the
behavioral dimension extending from extreme anxiety proneness on the
one hand to anxiety suppression and avoidance on the other. So far,
we have been exclusively concerned with inter-individual differences
in respiration and personality structure. Turning in the next section
to intra-individual differences, we will continue to focus upon the
behavioral dimension alluded to above.

Intra-individual differences in breathing.

Studies which may be classified under this heading are
numerous and go back to the early beginning of experimental psychology.
Several contributors to Wundt's _Philosophische Studien_ and to his
later _Psychologische Studien_ discussed this problem. In a historical
context, an important contribution was made by Lehmann's study in 1892
suggesting that pleasurable and unpleasurable situations have different
effects on respiration. Since that time various techniques have been
used to produce as realistic a situational impact as possible. Fear
has been induced by a sudden tilting of the subject's chair backward
into a nearly horizontal position, by sudden loud noises, by electric
shock, movies, etc.

Before going into the main results of these experimental
studies, we would like to inquire briefly into the respiratory changes
brought about by postural changes exclusively, by changes in the body's
position in relation to gravity. We want to start out with such a
broad perspective because so much of our reasoning so far has been

based upon the hypothesis that postural factors are of crucial importance in shaping an individual's respiratory pattern.

The question about postural factors influence on respiration
was alluded to in our previous reference to Wade's experimental findings.
Looking back on the table on page 31, summarizing his main findings,
we immediately see that differences exist in his subjects respiratory
pattern in an erect and supine position.

From our discussion of the anatomical basis of the respiratory
apparatus, more precisely, from our assertion that an extension of the
spine will raise the ribs, it follows that we should expect a somewhat
increased circumference of the chest in a supine posture (granted a
couch that is not too soft and really is stretching the back muscles),
as compared to an erect posture. This was found to be true. In the
erect posture the mean resting circumference of the chest was found
to be on the average, 16 mm. smaller than in a supine posture, although
the amplitudes of the chest movements were found to be very much the
same.

The greatest effect of postural change was interestingly
enough found with respect to the diaphragm. The total and tidal
diaphragmatic excursions were about the same, while the size for the
complemental and reserve movements were quite different in the two
postures. This coincides with similar differences in inspiratory and
expiratory ventilation capacities. In an erect posture, the resting
postition of the diaphragm was found to be on the average 39 mm. above
and 64 mm. below its maximal excursions, while in a supine position,
the same measures were 19 and 80 mm. respectively. Given that the
maximal excursions reaches very much the same positions in an erect
and supine posture, the measures indicate that the resting position
of the diaphragm in a supine posture is on the average about 20 mm.
above that in an erect posture.

In summary, Wade's empirical results show that postural
factors have great influence on the movement pattern of the diaphragm
and on its resting position, as well as on the resting position of
the circumference of the chest although not on its movement pattern.

Although one certainly ought to be extremely cautious in
generalizing from situationally induced to more permanently established,
-- genetically determined-postural features, Wade's data at least

suggest the considerable influence that might be brought to bear on
the respiratory pattern by muscular hypertonicity of an anti-gravita-
tioned nature, by skeletal muscular tensions being kept up always
and everywhere.

One of the main assumptions being made in the previous section
was that diaphragmatic excursions may be greatly restricted, in extreeme
cases possibly blocked and immobilized to such an extent that we find
an upward suction of the diaphragm during inspiration. This latter
viewpoint, suggested by Lowen in a couple of his case presentations,
has also been suggested by Weitz, Fleisch and others studying diaphragmatic movements fluoroscopically. A potential source of error in
these latter studies is the lack of control of the vertical movement
of the anterior attachment of the diaphragm. By controlling this
factor, Wade for instance, found that what may immediately look like
an upward suction or lack of movement of the diaphragm, by closer
examination is a descending movement being counteracted by a lifting
of the chest wall. Wade is probably correct in his pointing out that
diaphragmatic movements - in spite of many peoples subjective
belief - are outside ordinary voluntary control. However, such a lack
of control does not rule out the possibility that postural-tonic
factors may have a great influence on diaphragmatic movements. In fact,
Wade emphasizes strongly this very point although he limits their
influence to changes in the position of the body, i.e., to the varying
influence of gravity.

That tonic factors may operate on the diaphragm indepentent
of gravity changes is pointed out in studies by Faulkner (1941) and
Wolf (1947). For instance, Wolf maintains:

"Complaints of respiratory distress characterized by inability
to get full breath were found to occur commonly among anxious
individuals... By discussion of situational conflicts, attacks
were induced in 17 subjects during fluoroscopic observation.
In each instance a characteristic disorder of diaphragmatic
function was observed. Respiration became jerky and the
excursion of the diaphragm exeeded that in expiration. The
diaphragm thus assumed a progressively lower position. When
its contractile state was such that an adequate inspiratory
excursion was no longer possible, dyspnea occurred with a
feeling of inability to take a breath." (p. 1201)

Wolf's observation that the diaphragm descended more and more as its excursions were increasingly reduced, rules out any explanation in terms of an upward pull of the chest wall: In such a case we should have expected a progressively higher position of the diaphragm. To the extent diaphragmatic excursions are changed in response to the discussion of situational conflicts, it is likely to think that also more permanent changes may take place. Although we will not rule out the possibility that the diaphragm, existing in a state of hypertonicity, occasionally may be sucked up during respiration, we don't think this is by far a common feature of costal respiration, probably in most cases the diaphragm is only partially immobilized.

Wolf's observations of the diaphragm assuming a progressively lower and immobile position as conflictual situations were dealt with coincides with similar observations by Faulkner (1941), and does also support William's (1942) finding that children being confronted with a conflictual situation, a situation inducing opposing sets and attitudes, tend to react with a decrement in their abdominal breathing movements. Of course, it also fits well in with the different terms which have been used to cover situational anxiety attacks, terms like respirator corset, diaphragm-ring, etc.

The association between emotional stress and disorders of breathing, sighing with grief, gasping with rage, panting with fear, heaving with resentment and so on, has repeatedly been emphasized and utilized by poets and novelists. Psychologists, trying to demonstrate experimentally invariable relationships between breathing patterns and emotional states have been rather unsuccessful. Surveying the experimental literature on the subject, Woodworth (1938) writes:

> "There is agreement among experimenters that the breathing tends in excitement to be both fast and deep. This is the clearest known correlation between respiration and emotion. Pleasantness and unpleasantness show either increased or decreased breathing, usually increased, apparently because both pleasant and unpleasant stimuli are apt to be stimulating. Another clear correlation is that between momentary attention and partial or complete inhibition of breathing. Sudden stimuli will make the subject 'catch his breath'." (p. 260)

And more than a decade later, on the basis of a critical review of the literature up to that date, Clausen (1951) states:

"Although all results are not in agreement, the predominant
tendency observed seems to be that emotions increase the
beathing rate and increase the I/E ratio. A more doubtful
conclusion is that respiration becomes more irregular in
states of emotion. It seems, however, that there is not
sufficient evidence for the assumption that different
emotions have characteristic effects on respiration, in
the exception of the sudden inspiration in case of surprise."
(p. 15-16)

On the basis of our previous discussion of individual
differences in the thoracic-diaphragmatic respiratory balance, it
is not at all surprising that the results of various studies in this
area have not been in agreement, neither that one and the same study
have found large individual differences in the breathing pattern
accompanying the introduction of specific emotional stimuli. For
instance, from what has been said earlier, we would expect a person
with an anxiety prone respiration, that is, a person with a high
costal respiratory predominance, to react much stronger to activating
emotional stimuli than a person with a more flexible respiratory
pattern, and such a person again, to be more activated than a person
with a relatively rigid and fixated armor of the chest wall. That
this in fact is the case is strongly suggested by the results of
empirical studies by Finesinger (1944) and by Finesinger and Mazick
(1940a, 1940b), showing emotional stimuli to have significantly
different impact on respiration in various nosological groups. While
respiration was found to be altered considerably by emotional stimuli
in hysteria, phobia and anxiety neuroses, so it was found to be nearly
unchanged in hypochondriacs, reactive depression and compulsion
neurosis. While the former groups showed more pronounced changes in
respiratory regularity than was found in a normal control group, so
the reverse was true as regards the latter groups. We note of course,
in accordance with our previous reasoning, that the former groups are
proabably all characterized by a fairly high degree of thoracical
breathing, while the latter groups, most probably, are characterized
by a habitual diaphragmatic predominance.

An empirical study by Stevenson and Ripley (1952) also presents results supporting our present train of reasoning. Stevenson
and Ripley's study is focused upon the respiratory changes taking

place in 22 subjects as they were interviewed by a physician about various topics and attitudes known to be of relevance to their main life difficulties. The conversation about these topics and attitudes being separated by short periods in which the subjects were invited to relax and think about pleasant experiences. As subjects were used out-patients, fifteen with bronchial asthma and seven with anxiety states. In comparing the respiratory movements during relaxation with those during periods of emotional arousal, attention was paid to rate, amplitude, regularity, and the I/E ratio.

As we might have expected, great individual differences occurred on all of these variables: in 67%, the rate always increased during periods of unpleasant emotions, while in the remaining it sometimes remained unchanged or decreased; in 72%, the amplitude increased (in 42% always) during periods of emotional arousal, while in 28% it decreased (in 14% always) or remained unchanged; in 92% of the asthma patients (in 55% pronounced) the I/E ratio decreased during emotional arousal; while the same was true with 43 and 14% only of the anxiety patients -- the difference being highly significant; in 45%, irregularity was found invariably to occur during periods of unpleasant emotions; and finally, in 41%, sighing respiration was found to be a prominent feature.

What makes Stevenson and Ripley's study so interesting is that they do not stop by counting the number of subjects showing increased breathing rate, increased amplitude, etc. as an accompaniment of unpleasant emotions, but analyze their empirical material one step further as to what sort of emotions were involved, and how the subjects dealt with their emotional arousal. By such an analysis they arrive at a number of findings making apparent inconsistencies more explainable. In summary, they end up with the following conclusions:

1. Increased rate and/or thoracic amplitude and sighing is associated chiefly with anxiety, but also with anger and resentment when these feelings are openly expressed.

2. Decreased rate and/or thoracic amplitude is associated with a state in which the subject feels tense, reserved, and on guard, with unexpressed feelings of anxiety and anger, and with feelings of dejection and sadness.

3. Gross and remarked irregularity of respiration is associated with anger, particularly when suppressed (i.e. when feelings are mixed and ambivalence rather than clear hostility or anger predominate), and with feelings of guilt.

4. Prolongation of expiration is associated with situations of stress where decisive action is indicated but where the individual shows tentative, indecisive, evasive or withdrawing behavior. The higher proportion of patients with asthma than patients with anxiety exhibiting this trait is in accordance with indecisiveness and evasiveness commonly being a characterological feature of asthmatic patients.[1]

Stevenson and Ripley' study was focused upon respiratory variations in the same subjects over time, with no systematic attempt being made to analyse individual differences. However, it is tempting to assume that the intra-inividual variations found do correspond to characterological differences. For instance, their finding that increased thoracic amplitude corresponds to anxiety but also to other types of overt affect expressions, make it tempting to suggest that a potential thoracic mobility corresponds to an ability for affect expressions, and their findings that a decreased thoracic amplitude corresponds to reserve, guardedness and unexpressed feelings of anxiety and anger, tempting to suggest that a chronic thoracic immobilization corresponds to a permanent suppression and inhibition of affect expression. Of course, these suggestions stem from our theretical framework. Although Stevenson and Ripley's study did not prove the validity of these suggestions, their results certainly fit into the hypothetical formulations presented. They also substantiate the findings of a study by Borgen, cited by Haavardsholm (1946), that a "blown up" and fixated chest posture tends to subdue affective images and the expression of affects.

---

1) Also, prolongation of expirations is found in asthmatic attacks: "In some of the asthmatic patients the character of the pneumographic recordings changed markedly with almost perpendicular inspiration lines and long sloping expiration lines. In other tracings there was a plateau with delay before effective expiration."

Stevenson and Ripley's study comparing respiration during relaxation and emotional arousal indicates not only that emotional arousal is accompanied by respiratory changes, but also that the changes observed are different in different individuals and dependent upon the nature of the emotion aroused and the individual's habitual way of coping with the emotion in question, that is, dependent upon his character structure as expressed in his particular respiratory and postural pattern. Given such a viewpoint we should expect a closer association between emotional stimuli and respiratory changes in children than in adults, particularly in young children prior to the development and interference of distinct characterological patterns.

- - - - - -

We don't know about any study comparing the respiratory responses of children and adults to emotional stimuli, although it has been pointed out that the intra-individual variability of the respiratory pattern in children is much larger than in adults, and that the child's body involvement in breathing is much more total. For instance, Halverson (1941) has payed attention to the fact that the movements of an infant's head, hands and legs usually correspond closely to its respiratory movements in frequency and scope - - deep inspiration being accompanied by flexion of the legs and hands, indicating an increase in tension of the musculare of the whole body, and expiration paralleled by relaxation or extension of the limbs. As regards the greater variability in children, this is particularly manifested in the great changes found in their expiratory position from one situa-tion to another. In a study of 18 boys and 25 girls, ranging in age from 2 to 24 weeks, Halverson (1941) found changes in the thoracic expiratory position measuring more than twice the highest individual amplitude and more than 8 times the average costal amplitude, and changes in the abdominal expiratory position, measuring more than twice the highest individual amplitude and more than 6 times the average abdominal amplitude. These observations have to be compared with studies (e.g. Greene and Coggeshall's study cited by Halverson) showing undulations in the breathing curve of adults in various situations to be much more affected by inspiratory movements than changes in the expiratory position.

The fixation of the expiratory position of thorax and abdomen found in adults and the decrease in total bodily involvement in breathing, point to maturational differentiating processes, but raises also the question whether aging is the sole reason, whether developmental conflicts resulting in enduring postural fixations may not also in many cases represent an important contributing factor.

Granted the greater flexibility and responsiveness in children's breathing pattern, a close association should be expected between this pattern and their emotional arousal, ascertained from general behavioral observations. This is in fact one of the main conclusions drawn by Halverson in the study just referred to.

Describing his empirical data - consisting of simultaneous pneumographic recordings from chest (at the level of the nipples) and abdomen (over the umbilicus), taken in various situations (nursing, satiation-quiet, satiation--animated, satiation-going to sleep, early sleep, profound sleep, pleasureable activity, quiescence, restlessness, fretting and crying) -- Halverson notes "that each type of behavior represented in the various situations has its characteristic breathing pattern," and so characteristic did he find these situationally induced breathing patterns that he concludes rather definitely "that the physiological state of the infant is reflected in his breathing."

According to Halverson, distinct differences in respiration exist between profound sleep, early sleep, awake quiescence, restlessness, satiation-animated, pleasurable activity, fretting, crying; a dimension going from deep relaxation to strong excitement, a dimension covering, in other words, an arousal continuum. The differences between these states, Halverson indicates, can be measured and expressed quantitatively in relation to a number of respiratory variables. His main variables and conclusions are the following:

1. The respiratory rate: The breathing rate in infants increases with excitement and diminishes with quiescence and relaxation. The highest breathing rate occurs during crying (a mean of 133 respirations per minute was found here), while it is lowest during sleep (mean 32 respirations per minutes). In general, the longer the sleep, the lower the breathing rate.

2. The regularity of rate: Alterations in the rate of breathing increased from sleep to waking, and from quiescence to intense excitement.

3. <u>The I/E ratio:</u> During quiescence and sleep inspiration is usually much faster than expiration. Expiration is passive and is usually followed by an appreciable pause. When excitement increases moderately, the inspiratory and expiratory phases assume equal durations. When excitement increases still higher, the breathing movements show abrupt ascending and descending arches of varying heights and lengths. The breathing curve consequently becomes abound with spikes, dents and notches, and the relationship between inspiration and expiration, erratic.

4. <u>The regularity of the I/E ratio:</u> As just stated, the higher the excitement, the more erratic and irregular, the I/E ratio.

5. <u>The thoracic respiratory amplitude:</u> In general, the costal amplitude is higher, the higher the emotional excitement. The maximum mean costal amplitude are found during pleasurable activity. As excitement becomes very intense the thorax often tends to inflate and stiffen for brief or protracted periods of time. During fretting and crying the costal amplitude vary rapidly from one extreme to another. The movements at times are very large, at other times very small, so that in many instances the distribution of amplitudes are bimodal and the mean amplitude relatively moderate.

6. <u>The regularity of the thoracic amplitude:</u> As just stated, in the case of high excitement, there is a tendency for the thoracic amplitudes to show great variability. In general, the more relaxed the infant, the more uniform are the thoracic amplitudes.

7. <u>The abdominal respiratory amplitude:</u> The mean amplitude of the abdominal respiratory movements is greatest during profound sleep. The amplitude becomes somewhat smaller when the infant is quiet-awake. As excitement increases to high intensities, the amplitudes get larger again. However, in this case, the abdomen contracts and expands violently but intermittently also stiffens, so that the mean amplitude becomes relatively moderate.

8. <u>The regularity of the abdominal amplitude:</u> As just stated, in the case of high excitement, there is a tendency for the abdominal amplitudes to show great variability. In general, the more relaxed the infant, the more regular are the abdominal amplitudes.

9. <u>The thoracic-abdominal amplitude ratio:</u> The ratio between the thoracic amplitudes and the abdominal ones is generally greater, the greater the excitement. In strong excitement the costal amplitudes usually equals or exceeds the abdominal ones, while during sleep and quiescence, the abdominal amplitudes frequently are nearly twice the thoracic ones.

10. <u>The thoracic expiratory position:</u> The expiratory position of the chest is higher during waking than during sleep, and higher during excitement than during quiescence. In general, the higher the excitement, the higher is the expiratory position.

11. <u>The regularity of the thoracic expiratory position</u>: Alteration in the expiratory positions of the chest increases from sleep to wakening and from quiescence to intense emotions. Undulation in the breathing curve occurs in all situations except occasionally during deep sleep. Generally, the larger the undulation, the greater the excitement, and the more abrupt and angular the undulation, the more unpleasant the situation.

12. <u>The abdominal expiratory position</u>: The expiratory position of the abdomen is slightly higher during waking than during sleep, and tends to be higher, despite frequent low dips, during excitement than during quiescence.

13. <u>The regularity of the abdominal expiratory position</u>: As stated for the thoracic expiratory position, undulations in the breathing curve occur in all situations except occasionally during deep sleep. Generally, the larger the variability of the abdominal expiratory position, the greater is the infant's level of excitement.

14. <u>The thoracic-abdominal synchronization</u>: In deep relaxation, as found during profound sleep, an abdominal type of breathing prevails, that is, both inspiratory and expiratory movements are initiated by the abdomen, and a time lag occurs between the corresponding abdominal and costal respiratory movements. When the infant is relaxed awake, the respiratory movements are unisonal, that is, the movements of the chest and the abdomen are synchronized. As excitement increases, a costal type of breathing emerges, that is, movements begin with the chest and spread to the abdomen. When emotional excitement runs very high, antagonistic breathing movements tend to occur, that is, inspiration is affected by expansion of the chest and contraction of abdomen, and expiration, by the contraction of the chest and expansion of the abdomen.

15. <u>The thoracic cycle shape</u>: During quiescence and sleep the costal movements are usually of a skewed-hill-and-dale or a skewed-hill-and-dale-with-pause type. When excitement increases moderately, the movements assume a hill-and-dale type, and when it increases strongly, the movements show abrupt ascending and descending arches of varying highs and lengths so that no consistent cycle shape can be ascertained.

16. <u>The abdominal cycle shape</u>: During quiescence and sleep the abdominal movements are usually of a cusped or a cusped-with-pause type. When excitement increases moderately, here too the respiratory cycles usually assume a hill-and-dale type, and when the excitement increases strongly, here too no consistent cycle shape can be ascertained.

Although Halverson maintains that various behaviors and situations are accompanied by a characteristic breathing pattern, this does not imply that he suggests that no individual differences exist.

Quite the opposite, he states explicitly that his study "appears to
indicate that respiratory patterns differ among individuals and to
a great extent serve to identify them". And he continues:

> "No two of our breathing records were identical.... Differences
> between infants were observed in the rate and amplitude of the
> respirations, in the length and height of undulations, and in
> the time elapsing between inspiratory movements of abdomen and
> thorax." (p. 268)

However, in spite of these individual differences, by
comparing the respiratory curves of the same infant in various situations, he consistently finds that "the physiological state of the
infant is reflected in his breathing." Individual differences are
most clearly revealed during deep sleep and during quiet periods,
during periods when each infant's breathing cycles are relatively
uniform. On the opposite end, during periods of strong excitement,
the infant's breathing becomes inconsistent and characteristic
individual features tend to disappear. Often during such periods,
breathing becomes completely disorganized, that is, numerous alterations from one type of respiration to another take place, rapid and
irregular changes, shifts of an inspiratory or expiratory movement
from abdomen to thorax or vice versa, abrupt stops, marked differences
in amplitude of successive inspirations and expirations, temporary
suspension of movement by either abdomen or thorax, and so on.

The relationship noted by Halverson between excitement and
thoracic breathing and thoracic expansion fits well into our expectations derived from our previous discussion. And the same is the case
as regards the relationship noted between relaxation and abdominal
respirations and a low and regular breathing rate.

Perhaps one of the most interesting aspects of Halverson's
findings is the great sensitivity discovered as regards the infant's
expiratory position, that increase in excitement is shown by a rise
in the costal breathing curve, caused by a series of respirations
during which the volume of air inspired exceeds the amount expired,
and a diminution of excitement, by a drop in the costal curve, caused
by the excess of expired over inspired air in a subsequent series of
respirations. The very same phenomenon is observable when the infants
go from sleep to a waking state and vice versa, Halverson maintains.

Opening of the eyes after sleep usually occurs with the first of a
succession of very deep inspirations, and the transitional period
from waking to sleep with a successive decline in the baseline of
the respiratory curve, most pronounced in the case of the costal
curve, suggesting a particular sinking-into sleep pattern or phase.

Halverson's suggestion that respiration gives a pretty good
indication of an infant's activation or excitement level corresponds
to the fact that observations of respiratory changes are considered
to be one of the most reliable sources of information as regards the
level of anesthesia present in a patient during the administration
of an anesthetic drug.[1]

It also fits in with the fact that respiratory measures
have been found to be among the best discriminators in evaluating
information, i.e. in discriminating between lying and truth telling
in interrogation. Benussi's classical experimental finding was that
during lying the I/E ratio is increased because of a change in the
pattern of breathing. On the basis of recent laboratory tests.
Davis (1962) notes that both amplitude and breathing cycle time
(the inverse of rate) increase in both conditions, with the maximum
5 to 10 sec. after the delivery of a question. But in lying the
amplitude increases in a lesser amount and the cycle time in a greater
amount than in truth telling. The fact that breathing during deception
is shallower and slower should point in the direction of a lower
excitement level, but may also be interpreted as an overt manifestation
of the subject being more attentive, more emotionally and intellec-
tually controlled during deception.[2]

Of interest in the present context is also some observations
by Kubie (1953) derived from his experiments on hypnagogic reveries.
Describing how the amplified sound of the respiratory rhythm often may
exercise a striking hypnagogic effect, he states:

----

1) For instance, Clement (1951) discussing nitrous oxide as an
   anesthetic agent points out that a light plane of anesthesia
   usually is expressed in a slow, quiet, shallow and regular
   breathing, a profound plane, in a spasmodic, slow and irregular
   respiratory rhythm, and a normal surgical plane, in a regular,
   "machine-like" respiratory pattern.
2) This interpretation fits in elegantly with the results of
   Stevenson and Ripley's study previously referred to, cf. p. 65f.

"Sometimes the patient went into hypnagogic states with the evocation of highly charged memories and fantasies. At times he went into full hypnoidal states...as our subject entered into these different states, significant alterations occurred in the respiratory rhythm. Evidently the vagus complex and the ascending reticular substance may be closely allied." (p. 36)

Other investigations too have pointed out that respiration during daydreaming differs from the respiratory pattern found during directed thinking. For instance, Corwin and Barry (1940) note that daydreaming usually is accompanied by more shallow, slower and more plateaux-like breathing. Comparing respiration during sleep and hypnosis, Jennes and Wible (1937) note "a definite tendency for respiration rate and amplitude to decrease in sleep, but not in hypnosis." Furthermore, Paterson (1934) points out that respiration in a drowsy state contrasts to respiration during sleep; while in the former state, it is usually deep and slow, in the latter, usually slow and shallow. Jennes and Wible (1937) too suggest that "subjects tend during sleep to have slower respirations but shallow rather than deeper ones," while Bambridge and Menzus, cited by Haavardsholm (1946), claim that respiration during sleep not only is slower but also deeper. All the latter studies are based upon adult subjects. Again, we are faced with apparent inconsistencies, and again being faced with adults, we might suppose that characterological factors, factors shaping the impact of situational conditions, are of decisive importance, although the researchers lack of differentiation between various levels of sleep and various levels of hypnosis, is a most striking observation.

- - - - - -

To sum up our discussion in this chapter we want to reemphasize that breathing movements seems highly dependent upon characterological postural conditions. It is most probable, although so far not definitely proved, that the respiratory pattern may give valuable diagnostic information. While situational changes in infants seem to be transformed fairly directly into respiratory changes, so do adults exhibit a much more complicated relationship since situational

impacts always will be superimposed upon a more basic, characterological
respiratory pattern. In order to decipher the actual pattern found in
a given adult, a model has to be used accounting for the interactional
processes taking place between situational and characterological
factors. As regard the latter type of factors, our understanding
hinges upon our insight into developmental conditions, upon our ability
to construct a theoretical framework or model to account for individual
differences in postural patterns generally. In the next chapter we
will try to present a sketchy outline of such a developmental
posturological theory.

A DEVELOPMENTAL VIEWPOINT ON POSTURAL CONFIGURATIONS.

Sucking, biting, chewing, vomiting, micturition, defecation, crying, sneezing, orgasm, yawning, fear and anger, laughter and grief, may all be considered as basic organismic responses, responses having an innate biological source. They are all closely related to respiration. In fact, they may be considered so closely related that unless they become integrated with respiration their expression will neither be complete nor give full release and satiation.

As an introduction to our discussion of developmental aspects of postural configurations, we want to state the following six propositions:

1. The release of a basic organismic response presupposes a postural basis.

2. Emerging basic organismic responses will have a tendency to repeat themselves until they get integrated with breathing.

3. Their integration and coordination with breathing presupposes a learning process.

4. Only to the extent they get integrated with breathing will their expressions be complete and give full satiation.

5. Their release can be interfered with by a dissolution of their postural basis.

6. The permanent dissolution of the postural basis for an organismic response will have repercusions on the organism's breathing pattern.

Of course, far the greatest interest has traditionally been focused upon sucking, biting, defecation, and orgasm, organismic responses representing key stones in psychoanalytic developmental theory. We too want to discuss these responses most thoroughly, although before going into the area of psychosexual development we want to illustrate in a somewhat discursive way each of the propositions stated.

Release requires a postural basis.

This proposition should not need very much comment. It is
a common everyday experience that if we are not set for it, a joke
will not release laughter but boredom or embarrassment, an obstacle
not anger but bewilderment or surprise, a loss not grief but indif-
ference and a shrug of our shoulders. According to Braatöy (1954) a
fairly general precondition of the release of all sorts of emotional
responses is a certain openness, trust and contact:

> "Spontaneous emotional reaction is dependent on some minimal
> security or support. To really laugh you need a companion.
> To really cry you need another's shoulder or lap... To really
> express fear satisfactorily you also need a minimum of security...
> But support or security does not act immediately. The organism
> must first give up its drive to be prepared, stay on its own
> feet." (p. 186).

Following the last suggestion we may add, a rope dancer, for
instance, if he wants to survive, has to inhibit all spontaneous
basic responses while performing his art, he has to keep prepared and
stay on his feet. While not in the mortal situation, he may let go,
relax, open up. All human beings go through situations if not so
dramatic as rope dancing so similar to it in many respects. On the
other hand, not all human beings are able to oscillate between the
preparedness needed in these situations and the lack of preparedness
required in other situations. Their preparedness gets fixated
detrimental to their spontaneity and their ability to find the
minimal security and support to really laugh, cry, express fear and
other basic responses, in short, to be full human beings.

Unintegrated responses get repeated.

This proposition may also be illustrated by a quotation from
Braatöy (1954). Referring to the treatment of a woman of the anxiety
hysteric type, he states:

> "Working with her ad modum Jacobson, teaching her to tense the
> different muscle groups, to 'feel the tension' and then relax,
> the relaxation after some sessions called forth stretching and
> intense yawning... the relaxation was every time accompanied by
> a tremendous yawning, frequently dragging the whole body with
> it into a global, cat-like stretching-yawning movement. This
> repeated itself over and over again: In some hours more than

thirty times in half an hour... Sitting back and observing her, I suddenly saw that she did not yawn! Or, rather, a very important part of the complete stretch-and-yawn reaction was lacking, namely, the inspiration!" (p. 167).

In this case the physiological basis for the "repetition compulsion" is fairly clear. By not being complete, the yawning did not fulfill its essential purpose in bringing oxygen to the tissues, and by not changing the blood chemistry effectively, it left the organism in a "trigger state" for a new yawning; the next yawning also being incomplete did not change this basis, and so the yawnings were repeated again and again. Although the physiological basis is less clear, the mechanism behind fixations, i.e. repetition tendencies generally, is probably very much the same. A complete sexual orgasm leads to relaxation and satiation, a fragmentary, incomplete orgasm, to a state of irritation, lack of satisfaction, that is, to a "trigger state" - in some individuals, accordingly, to sexual hyperactivity. A complete expression of grief and crying, to relief and relaxation, a fragmentary expression, to a lasting depression. A complete expression of fear, to tranquility, a fragmentary expression, to nervousness or neurotic symptoms. A complete expression of laughter, to a feeling of well being, an incomplete expression, to repetitive giggling. A complete expression of anger, to relief, an incomplete expression, to lasting irritability or temper tantrums, And so we may go on and describe different socio-biological reactions.

Integration requires learning.

To illustrate this proposition we will turn to the results of a study by Halverson (1944) focused on the relationship between sucking and breathing in infants from birth to the age of 24 weeks. In this study Halverson observed that the chest and the abdomen of infants function not only in breathing, but also in reinforcement of sucking movements, the chest by expanding and the abdomen by contracting. The effect on breathing of these sucking-reinforcing movements is to increase the costal movements and to inhibit the abdominal breathing movements. However, the sucking-reinforcing movements do not necessarily coincide with the breathing movements. In fact, Halverson observed that the age at which coordination between

breathing, sucking and swallowing was first exhibited varied extensively from infant to infant. One infant showed good coordination
eight hours after birth at her first feeding. Three infants showed
coordination within the first three days while two of his infants
still consistently failed to obtain a steady temporal relationship
between breathing and sucking at the age of 24 weeks, at the time
the study was stopped. The one of the two infants is described as
"appeared always to be hungry and become excited at feeding time",
the other as "never appeared to be hungry, and was easily disturbed
when she did feed".

Focusing on the factors probably responsible for inadequate
coordination, Halverson points to the following ones:

1. The adaptability of the infant, i.e., the ability of the infant
   to adjust his suction power to the pressure required to obtain
   the milk in an appropriate amount so that swallowing may occur
   at regular intervals. It seems that regulation of the suction
   power and timing of its output rather than sheer power itself,
   are the crucial factors. In fact, no relationship was found
   between sucking power (predominance of costal breathing) and
   consistent coordination.

2. The shape of the nipple, i.e., the nipple may be obstinate, or
   it may yield the milk too easily. Coordination never occurs
   when an infant is forced to exert himself unduly to obtain milk.
   In such a case, sucking cannot be maintained uniformily in power,
   and breathing is brought to a stop for each suck. Postural
   changes, stiffening, and abrupt bodily movements come into play
   in order to reinforce the sucking movements. On the other hand,
   a nipple that gives food too easily results in a situation in
   which the milk entering the mouth per suck becomes too great in
   amount to be disposed of adequately by an infant's normal manner
   of swallowing. Consequently, interference with breathing occurs.

Looking at these two conditions, the quality of the nipple
is certainly an external situational factor, while the adaptability
of the infant refers to more organismic conditions, although not
necessarily of an hereditary origin. Halverson points out that both

unusual avidity in feeding, restlessness, and rapid breathing (the
breathing rate exceeding 80 respirations per minute) seem to interfere
with the infant's adaptability, while "coordination was characteristi-
cally good when subjects appeared relaxed and at ease in the feeding
situation..."

Halverson's observations strongly indicate that an early
respiratory learning process is involved in the feeding situation,
that avidity, restlessness, etc. impede this process, while it is
furthered by relaxation and an appropriate shape of the nipple.

Our proposition implies that very much the same type of
learning process as that suggested above for sucking and breathing,
is involved in relation to yawning, biting, anger, and so on.
For instance, in expressing anger many people stop breathing or
largely subdue it. The close connection between breathing and
delivery pains has been specifically stressed by Read and others
advocating "childbirth without fear" and special training programs
for expecting mothers.

E.g. Kitzinger (1962) writes:

"The way a woman breathes is closely connected with the rhythm
to which her body adapts itself during the process of labour.
If she succeeds in harmonising her breathing with the contractions
of the uterus, which have a definite rhythm of their own and are
like waves in the way that they gather, rise to a crest and then
die away, she will be able to keep control of her labour and
instead of its being a muddle of painful sensations she will
maintain conscious control and find it very exhilarating."

The fact that both training courses and eductional literature
exist in this area emphasizes that integration and "harmonising"
evidently requires learning. Referring to observations of monkeys it
is not unlikely that full sexual orgasm also requires learning. Here
too we are probably confronted with a learning process related to the
integration of breathing and pelvic motor responses (Reich 1942,
Feldenkrais, 1949).

Satiation requires integration.

In order to illustrate the proposition we could continue
quoting Kitzinger to the effect that by synchronizing the breathing
with the rhythm of the contractions it is possible to make labor
pleasurable, and the energy-release involved in child-birth, a deeply
satisfying experience. However, to focus on a more everyday response
pattern we may again turn to a quotation from Braatöy (1954):

"In general practice it happens that the arrival of the physician sets off a tremendous crying in the child, which is a bit surprising to the family because the child tries in other ways to cooperate. The child is not simply panicky or stubborn. Nevertheless, it seems impossible to stop the crying and get on with the examination. If one tries to comfort the child, the crying or wailing increases in a tremendous crescendo; and if one tries the opposite attitude and talks in a "military" fashion, "But this is nonsense, Peter, you have been much braver in more dangerous situations!", the child screams as if he was being murdered. In such exasperating situations, I have sometimes succeeded by approaching the child from the point of view of the mechanics of emotional display. In a non-committal voice I ask the child to look at me. If he is able to do this I ask him "to breathe with me." Face to face then, I breathe in the rhythm of the child in a clonic sobbing way, only a little less accentuated. By following the child's respiration in this way, I make it possible for the child to follow mine. If I succeed in this, I gradually but quickly change my breathing to ordinary deep respiration. Following my breathing with some catching interruptions the child to its own surprise gets out of the emotional storm" (p. 180).

And a little later on he continues:

"In the pseudotemper tantrum described above, the incomplete reaction was connected with a partial blocking of respiration. This blocking exasperates the child and makes him further irri- tated and inclined to cry of protest. By helping the patient to breathe, one may release the final deep sob which leads up to the physiological ebbing of the emotional storm. Maybe most socalled temper tantrums are in this way pseudotantrums. They are released for nothing out of a trigger state and repeat them- selves because a postural (attitudinal) set blocks the complete physiological release. This comparatively rigid set induces at the same time a feed-back until the reaction is either tired out for the time being or the organism, by chance or with help, succeeds in performing the little shift which permits the complete reaction leading to satisfaction and rest." (p. 181).

Following Braatøy's train of reasoning, a postural state was producing the basis for crying but preventing "the final deep sob", i.e. the integration of crying and breathing. We may ask why did the child assume such a posture? Why didn't an integration take place? The answer to both of these questions probably hinges upon the child's earlier experiences, his experiences not providing the necessary conditions for a learning process to take place. Not having firmly established an integration between crying and breathing, the crying will probably have a tendency to repeat it- self. That is to say, only by such an integration will an organism eventually be able to outgrow the perpetuating influence of many infantile impulses and responses.

Inhibition requires postural changes.

This proposition parallels the first one. In the same way as the release of an organismic response presupposes a postural basis, so does a temporary or chronic inhibition of a response also require a postural foundation. In the above example of a boy with temper tantrum, it was implied that postures may prevent as well as further an integration process, and that only by a modification of the young man's posture and respiratory pattern he was able to get release of his crying spell. However, another solution might also have been possible.

"Big boys don't cry," "girls don't bite," "but this is nonsense, Peter, you have been much braver in more dangerous situations" are examples of parental pressure brought to bear on children in order to get them to relinquish infantile responses. Very often parents succeed in changing their childrens' behavior by such pressure - that is to say, they succeed in forcing their children to adapt to the pressure, not through processes of integration and learning but through changes in their postural patterns more or less permanently inhibiting the expression of the objectionable responses. Instead of the responses being outgrown, we would in such an instance talk about responses being repressed and fixated in a latent form.

What has just been said doesn't mean that all chronic postural changes are brought about in a purposeful way by parental interference. Every child has to pass through phases in which it has to adapt not only to its environment but also to maturational processes taking place within itself. In so far as the child is not interfered with but neither met and helped and supported by its environment, its newly emerging responses, needs and approaches, will have no possibility of taking root, of being integrated with breathing. Consequently they will be drained off, but this draining off process too probably implies postural modifications.

The above train of reasoning parallels very much the distinction being drawn between developmental and defense-inducing child rearing practices.

Postural dissolutions have respiratory consequences.

This last proposition follows from our assumption that a postural dissolution of the reaction basis of a given response represents a modification of the tonicity of skeletal muscles, of muscles related to the response in question. Since an individual's respiratory pattern, broadly speaking, is influenced by the tonicity not only of the diaphragm and the intercostal muscles but by all skeletal muscles·of the body, a postural dissolution will immediately have repercussions on the respiratory pattern.

This latter inference confronts us with a particular phenomenon. In the first place we have been suggesting that an integration and coordination has to take place between respiration and an organismic response in order for the latter to give full satiation. Now we are assuming that the dissolution of an organismic response may have repercussion on respiration, on its flexibility, restricting its adaptability in relation to other responses. In essence, what we are suggesting is that an early developmental conflict in a child's life resulting in a defensive solution will reduce the child's capacity to cope with later conflicts. In this respect we are of course reminded of Freud's developmental model comparing the libidinal development in children to an army advancing on enemy territory, an army that has to leave behind occupying troops at all exposed points, troops and forces that reduce its strength available for further advance. Focusing our attention on the respiratory profile, whether it is predominantly thoracic or abdominal, whether deep or shallow, whether the chest and the abdominal walls are kept high or low, etc., we extract information pertinent to the muscular conditions not only of the chest and of the abdomen, but indirectly also of the shoulders and the pelvis, the arms and the legs. It has been said that retracted shoulders convey repressed anger, a holding back of an impulse to strike, that raised shoulders are related to fear, that square shoulders express a manly attitude of shouldering one's responsibilities, that bowed shoulders convey the sense of burden, etc. In all these cases we should expect the thoracic motility to be influenced in one way or another. For instance, retracted and square shoulders we would expect frequently to coincide with a "blown up" and

immobilized chest, and raised and bowed shoulders, by and large,
with a much greater amount of thoracic motility. Striking
similarities exist when we focus on the structure of pelvic tensions
and its relation to abdominal motility. The pelvis may be free
swinging (giving the individual grace in gait and movement), or it
may be immobilized in a backward or forward position.

      Compared to the shoulder girdle, the pelvis, the jaw
(the jaw might also be free or immobilized in a retracted or protruded position), the eyes, the legs, and so on, an individual's
respiratory pattern represents a much higher degree of complexity
in terms of neural innervations, as well as in terms of the multiplicity of potential, functional derivations. It is reasonable
to think that the respiratory pattern being the resultant of a
multiplicity of factors will correspond to psychological traits
of a fairly generalized nature. Its very complexity of determination, probably provides us with an entrance to higher-order personality characteristics. On the other hand, this very complexity
prevents us from an analysis of all the part processes and developmental fragments that go into the formation of a given personality.
To be able to understand and describe a personality we would have
to go into these fragments and processes, that is, our diagnostic
procedure would have to transgress the realm of respiration and
enter into the area of posturology proper. Entering into this area
we would be in need of a sort of theoretical model, a model making
explicit not only some of the major principles governing personality
formation - principles like those described earlier in this chapter -
but also linking various postural manifestations content-wise to
specific impulse-defense configurations. This latter aspect requires a developmental viewpoint, the spelling out of the postural
basis for and the nature of various impulse patterns emerging in
childhood. It requires, in other words, our entering into the area
customarily referred to as psychosexual developmental phases. In
order to establish a certain perspective for our own discussions of
these phases let ut start out with a brief summary of the
psychoanalytic viewpoint.

Psychoanalytic conceptions of psychosexual development.

Psychosexual development represents an integral part of classical psychoanalytic thinking. To illustrate this position more clearly we will describe briefly some central features in the writings of Freud, Abraham and Fenichel.

According to these authors, basic human motives (instincts) can be classified into two groups: 1) Those dealing with physical needs, e.g. hunger, thirst, defecation and urination, and 2) those dealing with sexual urges. Both types of motives are assumed to function from birth onwards and can be described by the following three characteristics: 1) aim, 2) object, and 3) source.

In contrast to physical motives, sexual motives are supposed to have the capacity to change, to alter aims and objects, to disappear from consciousness and to reappear in different disguises. The energy of sexual motives is called libido and in each individual a certain quantity of libido is assumed to be present from birth. In the course of an individual's development, libido becomes attached to various bodily zones or organs and undergoes a variety of trans- formations, e.g. from self to other persons, from progression to re- gression, from fixation to sublimation, etc.

Infantile sexuality differs from mature sexuality in three ways: 1) The areas which afford greatest sexual pleasure, 2) the aims of sexual urges, and 3) the objects through which the sexual urges reach satisfaction.

In Figure I (on the next page) is presented a schematic survey of this developmental model. The designations of the various stages refer to the area or zone affording greatest sensitivity. The developmental process, the investment of libido, is considered to move from the oral to the anal to the genital zone.

Briefly, we may say that the very first sexual aims are assumed to be of a pre-ambivalent, oral erotic character, i.e., sexual satisfaction is thought to be brought about by a discharge which dispels a condition of excitement in the oral zone, and furthermore, this discharge is thought to imply a certain mode of reacting. Later on, it is assumed, that the child goes through a second oral stage and two anal stages where an ambivalent orienta- tion and sadistic impulses represent partial drives and a normal

Figure I

Schematic Survey of Classical Psychoanalytic Views Regarding
Pregenital Developmental Stages

Stage	Sexual aim	Sexual object
Earlier oral	Preambivalent, autoerotic: Pleasure in sucking and oral incorporation (oral erotism)	Own body (without object). Situationally determined objects; milk, breast, bottle (anaclitic pre-objects)
Later oral	Ambivalent, narcissistic: Pleasure in biting and total (destructive) in-corporation (oral sadism or cannibalism)	Mother and mother-substitutes (narcis-sistic object)
Earlier anal	Ambivalent, partial love: Pleasure in expelling and destructive elimina-tion (anal sadism or expulsion)	Excrement and persons involved in the child's toilet training (partly narcissistic, partly altruistic object)
Later anal	Ambivalent, partial love: Pleasure in retaining and destructive control-ling (anal sadism or retention)	Excrement and persons involved in child's toilet training (partly altruistic, partly narcissistic object)
Earlier genital	Ambivalent, partial object love: Pleasure in masturba-tion, exhibitionism and destructive penetration (castration) (phallic-urethral sadism)	Parent of the opposite sex (incestuous object)

mode of reacting. These modes are also considered central in the phallic or early genital stage which follows and forms the transition to the post-ambivalent, genital primacy of the fully matured personality.

Changing our attention from the orthodox psychoanalytic model to Erikson's (1950) conceptions, there are a couple of features which strike us as being different. Erikson does not specify sexual aims and objects, but concentrates exclusively on what he calls organ modes or general modes of approach; he does not differentiate between an earlier and a later anal stage, but between the infantile genital stage in boys and girls; he does not consider sadism and destructiveness an innate characteristic of pregenital stages, and consequently he does not consider repression, fear of punishment, and fear of loss of love, a prerequisite for a child's educability and socialization which is the logical consequence of the orthodox psychoanalytic model. Figure II presents a schematic survey of Erikson's point of view regarding psychosexual development (See next page).

Erikson's point of departure is to some extent the same as that of the classical view, i.e. in each child different bodily zones successively become the focus of psychosexual functioning. Erikson differentiates the following three zones: 1) "oral-sensory", which includes the facial apertures and the upper nutritional organs, 2) "anal", the excremental organs, and 3) the genitalia. In each stage of development, Erikson maintains, a certain mode of approach will emerge and dominate the whole organism and its relation to its environment. Given a mutual regulation between the child and its mother in the first oral stage, an incorporative mode will dominate not only the oral zone, but become generalized to all sensitive zones on the body surface. In the same way, given a mutual regulation between the child and its surroundings, in the muscular, anal-urethral stage, the retentive and eliminative modes will from their focus in the anal zone become generalized to the whole of the developing organism, to its muscular as well as its social spheres.

Figure II

Schematic Survey of Erikson's View Regarding Infantile
Psychosexual Stages

Stage	Dominant Mode	Social Modality	Nuclear Conflict
Earlier oral-respiratory-sensory	Passive incorporative	Getting	Trust vs. mistrust
Later oral-sensory	Active incorporative	Taking and holding on to	
Muscular anal-urethral	Retentive-eliminative	Letting go and holding on	Autonomy vs. shame, doubt
Locomotor infantile-genital (male)	Intrusive	Making, i.e. being on the make	Initiative vs. guilt
Locomotor infantile-genital (female)	Incorporative (inclusive)		

A mutual regulation between the child and its surroundings
can be disturbed either by the child itself or by its significant
object. This being the case, the child may develop an inappropriate
mode, or more correctly, a previous auxiliary mode of approach may
grow into dominance and thus in a permanent way disturb the child's
psychosexual functioning. Theoretically Erikson considers it possible
for each one of the five modes he describes, i.e. passive incorporative, active incorporative, retentive, eliminative, and intrusive,
to become dominant in relation to each stage. The dominance of an
inappropriate mode, will interfere with the child's learning of
basic modalities. In the first oral phase, given a mutual regulation between the infant and the maternal source of supply, the child
learns to get, to receive and take what is given, and to get others
to give what it wants. By this learning "to get", the necessary ego
groundwork develops for the baby's capacity "to give", Erikson
maintains. Thus, what Erikson designates as social modalities may
be looked upon as a type of ego capacities.

Erikson deals at length with the relationship between
psychosexual stages and ego development. At each stage the child has
to adjust itself to the external world, and to the biological changes
which take place in itself. In the early oral stage the child has to
learn to regulate its organ system in accordance with the way in
which the maternal environment organizes its methods of child care;
in the late oral stage, it has to learn to combine sucking without
biting or to untie its unity with the maternal matrix, and in the
anal stage, it has to learn to coordinate tendencies towards retention and elimination. The various stages represent "problems" which
have to be solved. The solutions arrived at will influence the
child's total development and its enduring orientation throughout
life. The various stages may be considered as nuclear conflicts
for the child's growing ego formation. By successfully adapting
itself to the changes taking place, and this, of course, to a very
high extent depends upon the emotional support received from the
environment, the child's ego will not only acquire specific capacities, but also specific qualities.

According to Erikson, the enduring pattern of solution
established in the oral stage will be decisive as to whether basic
trust or basic mistrust is to characterize the ego. The consequence of the anal phase and the solution the child then adopts
will be decisive as to whether autonomy or shame and doubt are to
become a quality of the growing ego. Further, Erikson regards the
infantile, genital phase as decisive for the dimension "initiative -
guilt", the latency period as decisive for the dimension "industry -
inferiority", puberty and adolescence for "identity - role confusion",
the early adult years for "intimacy - isolation", and adulthood for
"generativity - stagnation" He ends his account of the different
phases of life with what he calls "maturity". Here "integrity" as
opposed to "disgust and despair", stands out as the keystone of the
developmental cycle.

After this brief introduction we will turn our attention
to a description of the various phases of development. With Abraham
we shall distinguish between five pregenital phases; the early oral,
the late oral, the early anal, the late anal, and the early genital
phase. A detailed description will be given of each of these phases
in accordance with our desire to present a survey of the impulse
patterns and modalities we think are most characteristic. In so
doing we will deviate somewhat both from the classic psychoanalytical,
as well as from Erikson's point of view.

Since one of our main propositions involves the close
connection between postural development and respiration, a brief
review of the respiratory development in children may serve as a
suitable starting point.

Respiration in children.

The child in utero receives its oxygen and discharges its
carbon dioxide through the placenta. Although the lungs at this
time contain no air, it has been pointed out by various researchers
that respiratory movements of several minutes duration occur in the
human fetus. Rhythmic but irregular movements varying in number
from 40 to 80 per minute have been attributed to contractions of
the respiratory muscles. It has been assumed (Feldman, 1920) that

the purpose of these movements is to strenghten the diaphragm and
thoracic muscles for use after birth. However, it has also been
assumed (Ribble, 1944) that the diaphragmatic contractions in
fetal life may function as a sucking apparatus supporting the heart
in pumpihg oxygenized blood from the placenta.

The lungs of the neonate are relatively small. However,
their growth is extremely rapid. At three months their weight is
about double that at birth, and at 11 months about double that at
three months. The structure of the thorax of infants differs in
various ways from that found in later life. Firstly, it is generally
somewhat conical in form with its smaller circumference at the level
of the arm pits. In adults the thorax is more barrel-like in shape.
Secondly, the anterior-posterior diameter is equal to, or even
greater than, the transverse diameter. At puberty, the transverse
diameter generally is more than double the anterior-posterior diameter. Thirdly, the position of the ribs in the neonate extends
almost at right angles to the spinal vertebrae. As the child grows
older they gradually sink and in adults show a quite pronounced
obliquity.

The breathing of infants has been described as diaphragmatic in type. Partly this has been considered due to the horizontal
posture, partly to the weakness of the bones and the muscles of the
thorax. As a result of intestinal pressure against the lungs,
breathing is shallow and an adequate ventilation thus can be provided
only by a fairly high frequency of respiration. The breathing rate
at birth has been reported to be between 30 and 80 respirations per
minute; at one year, between 25 to 40 respirations per minute; and
in adults, usually between 10 and 15 respirations per minute.

When the child assumes an erect posture, the consequent
sinking of the abdominal organs and anterior thoracic wall, and the
increasing downward slope of the ribs, are conducive to deeper
breathing. With the gradual strengthening of the thoracic muscles,
the movements of the chest become larger. After the third year
breathing will be more thoracic, and after ten years, the child's
breathing movements will rapidly assume the characteristics of
adults.

It has been suggested (Neilson and Roth, 1929) that some
types of respirations are hereditarily determined. Other researchers, stressing the high frequency, irregularity and easy irritability
of breathing in early infancy, have questioned the completeness of
the development of stabilizing mechanisims in infants and thereby
suggested the heavy dependency upon external factors and learning
in shaping the infants breathing pattern. As previously mentioned,
our own view coincides with this latter assumption, although we
would also like to emphasize Halverson's statement, quoted in an
earlier chapter, that the respiratory pattern differs among infants
and to a great extent serves to identify them.

The early oral phase of development.

A newborn child is completely helpless and its existence
is at the mercy of the sympathy and care of the persons around it.
A great part of the child's day is spent in sleep.

At birth the vocal organs are only rudimentarily developed,
but after a few weeks the nervous system has linked up with the vocal
chords, and after a few months it is possible to differentiate
between different sounds made by the baby. Some will be fretful
whimpering or weeping; others, such as crooning, gurgling, belching
and smacking its lips, may make us think of more pleasant experiences.
In certain situations we will observe a lively organism jerkily
sprawling and kicking its limbs as if it lacked a central organ to
coordinate and control them. In other situations the picture may
change completely. When the infant is being suckled we will usually
discover a finely attuned coordination of the muscles in and around
the mouth and throat. If we follow the child further we will often
notice that it is far from satisfied by its sucking of the nipple or
teat. It has a tendency to suck anything within reach, preferably
round, soft objects. Besides its mouth the surface of its body, and
the areas around the nose and the forehead in particular, seem to be
central instruments of pleasure. Stimulation in the form of fondling
and caressing gives rise to signs of pleasure. Although the mouth is
the main window through which the child achieves contact with its
environment, it may be said that its whole body takes part in sucking

in external impressions. It is true that the infant is completely
dependent on the care of others, but it is not merely passive and
receptive. Searching movements can already be observed at birth,
particularly in connection with feeding and later in connection with
sight and hearing. The first searching movements seem to be
reflexive and do not imply any intention on the part of the child.
Gradually the reflexive basis grows weaker and the child shows its
first genuine signs of social orientation. From the beginning the
infant displays greater bodily ability to make bending rather than
stretching movements, as if it were more ready to receive and take
in than to give; but along with the receiving there, nevertheless,
occurs very early - as far as oral activity is concerned - a significant element of seeking and approaching, so that even at birth,
in our opinion, one can speak of a basis in the child itself for
active emotional interaction with its environment.

The mouth and the surrounding area make up the dominant
system of organs in the infant's life economy. One can almost say
that initially the child's life is organized around its mouth. It
takes in impressions, pleasant and unpleasant, through its mouth;
it expresses itself, whether delighted or displeased, through its
mouth. The child's first vague perception of its environment seems
to be based on the experience gained by its mouth. Its mother will,
therefore, play a decisive role in its first social interaction.
A commonly held view is that out of this interaction will grow an
affective foundation, a foundation that will determine to a substantial degree the child's subsequent affective attitude to its
environment. If its mother's breast is experienced as pleasure-
giving, if the baby's sucking is not interfered with by the fact
that too much or too little effort is needed in acquiring milk, and
if suckling takes place in a timeless and harmonious atmosphere with
the mother's arm and body holding and supporting the child and
giving relaxed, warm contact, a germ of trust, confidence and opti-
mism will be formed which will influence the child's further
interaction with its environment, and support its initial reaching
out and its emerging social orientation.

Granted that sufficient emotional support is offered by
the external world, the development of searching and responding
impulses will assume a central position in the infant's maturation
process. These searching and responding impulses will have the
form of pleasure-getting avenues. The early oral phase may be
considered as the stage where these impulses crystallize, as the
stage where a <u>receptive responding modality</u> emerges.

In our discussions so far we have relied heavily upon
Halverson's observations indicating that a learning process has to
take place in order for sucking to be fully gratifying, and that
the learning process in question is largely dependent upon the
child's total situation.

If no learning process takes place whether due to the
fact that the supply of milk is insufficient, more milk is forced
into the baby's mouth than its digestive system will accept, or
feeding is rigidly bound to a time table out of step with the
infant's own rhythm or pursued in an atmosphere of restlessness,
irritation and rejection, we are faced with various possibilities.
One of these possibilities is that the sucking response will
continue to assert itself in different ways. But the possibility
also exists that the constantly ungratifying situation will give
rise to postural changes in the infant, changes preventing its
repitious frustrations.

From our earlier discussion of Halverson's observations
we recall that one of his two subjects who consistently failed to
obtain a steady integration between breathing and sucking at the
age of 24 weeks, at the time the study was stopped, was described
as "appeared always to be hungry and became excited at feeding
time", and the other one as "never appeared to be hungry, and was
easily disturbed when she was fed." The former infant exemplifies
a repetitious tendency, the second one a more defensive position.

That postural defenses may emerge at a very early age as
a result of oral frustrations is indicated by Ribble's (1944) observations. On the basis of extensive observations of infants'
responses to what she calls inadequate mothering, she differentiates

between two types of behavioral syndromes, negativism and regression, both characterized by changes in the infants' very relationships to sucking.

By negativism Ribble refers to the development of a negativistic excitement in relations to the feeding situation, accompanied by a posture characterized by the tensing of the deep muscles of the neck, of the back of the neck and the back, giving the torso a slightly backward arch, and by the stiffening and resistance of extension of the arms and the legs. Periods of violent crying occur in which the child is impossible to calm down. Likewise vomiting and periods of prolonged restlessness. The child's breathing is described as shallow with tendencies to temporary suspension of breathing.

By regression Ribble refers to the development of an increasing passivity and apathy with a weakening of the sucking reflex and with sleep occupying a disproportionate part of the child's day. Feeding is immediately succeeded by sleep lasting until the child is awakened for its next meal. In extreme cases, the sucking reflex may disappear altogether, and the infant's sleep gets the quality of stupor or coma with a shallow, irregular, abdominal pattern of breathing, suggesting an almost complete regression to a prenatal level of functioning. A most characteristic postural trait is a fairly generalized hypotonicity.

The two reaction patterns just mentioned may probably be considered prototypes of postural defenses - the one going in the direction of passive withdrawal, regression and hypotonicity, the other in the direction of protest, active repression and hypertonicity. We may think of these two patterns as possible solutions to each and every basic response being consistently frustrated, either actively by parental interference, or passively by not being responded to at all.

It is interesting to note that Ribble pays special attention to respiratory phenomena. She points out that the breathing of infants is precarious for probably the first six months of life, and that an adequate respiration during this period is of vital importance to the metabolism of the developing brain tissue. She also emphasizes that an association seems to exist between sucking and respiration,

the act of sucking having a reflex effect in stimulating respiration.
In the case of inadequate sucking and sensory stimulation, a
regression may take place. Regarding the respiratory aspect of this
regression she has the following to say:

"It has been shown that definite breathing movements occur in
the fetus - these movements are related to the muscles of the
diaphragm, which in the fetus, arches high into the chest
cavity over the large liver like a suction cup. It is thought
that in all probability its action is to suck oxygenated blood
toward the heart and chest generally from the region of the
placenta. After birth when air is inspired into the lungs, this
action becomes reversed. However, until the fetal circulation
is obliterated and the innervations of the diaphragm are well
developed, it appears possible for the organism to revert to the
prenatal form fo breathing activity... As a consequence of the
underdeveloped state of the central nervous system, this prenatal
or splanchnic orientation of the infant may persist, and in all
probability it is the biological basis for the tendency toward
regression and acute inanition." (p. 636).

Somewhat similar thoughts on the possibility of respiratory
regressions have been presented by Kempf (1930) in his discussion of
the somatic mechanisms behind the introverted affective adaptation
found in catatonics:

"All the autonomic segments or mentally essential organs are,
in a functional sense, completely developed at birth and do not
go through any marked changes except the lungs. Upon birth a
tremendous functional change must take place in the lungs
(respiration) to meet the struggle for life, and it appears that
this is the first visceral segment to regress under too severe
punishment. The organic cause of the susceptibility for
respiratory regression has been overlooked..." (p. 184).

Turning back to Ribble's description of negativism as a
state characterized by periods of violent crying in which the child
is impossible to calm down, we may suggest that the crying in this
instance has a repetitious quality, a quality somewhat similar to
the repetitiousness earlier described in relation to sucking
behavior.

Besides sucking, crying is perhaps the most dominant
expressive avenue of the infant. Here too a learning process, an
integration with breathing, probably has to take place in order
for the response to be complete and relieving.

Until the infant is about three months old it has no possibility for active refusal, by closing its mouth or shaking its head. During the first half of a child's first year of life its protest and rage get mainly expressed through vomiting and crying.

Discussing crying in early childhood we would ultimately have to distinguish between various forms, between the lusty cry of enraged healthy infants, the violent, perpetuating cry of infants in a state of negativism, and the clinging, sustained and stacato-like cry of infants in grief and severe disappointments.

Far from all severely orally deprived children in the first months of life regress into what Ribble calls a splanchnic state. The more congenital the child, the more it will protest, struggle, toss the body about, jerk up and down, move from side to side, and cry when deprived and frustrated. But, as stated, in the same way as sucking to be fully gratifying depends upon a certain external emotional support, so does crying also depend upon such a support.

Describing the breathing patterns of infants while crying, Halverson (1941) states:

"... crying usually begins while the lungs are well inflated... in prolonged cries... the air is forced from the inflated lungs in a series of explosive, expiratory movements interrupted by minor respiratory movements and catches of the breath... The costal curve frequently remains at a high elevation... Hence costal respiratory movements at this time are necessarily small. The movements of the abdomen are larger and more violent... Each cry usually starts on a major down stroke of the abdominal curve." (pp. 283-84).

In short, breathing during crying is usually characterized by a deep, quick inspiration and a smooth or interrupted prolonged expiration; by rapidity of breathing; and, in contrast to quiet breathing, by no pause between respirations. What interests us here, however, is Halverson's observation that breathing during strong excitement, during violent crying, may become disorganized. The sequence of the respiratory movements of thorax and abdomen during such periods defies accurate description, Halverson maintains.

Not only do numerous alterations from one type of respiration to
another occur but the irregularity of the changes, the shift of
an inspiratory or expiratory movement from abdomen to thorax, the
abrupt stops, the marked differences in amplitude of successive
inspirations and expirations, the temporary suspension of movements,
attest to a breakdown of organized action of the respiratory
mechanism.

Unless the infant is able to "keep breathing", no integration with crying responses will take place. Since too much excitement disorganizes breathing, the learning process in question seems
to be furthered by the extent to which the infant is prevented from
too traumatic or too upsetting experiences, and is clamed down when
a too strong excitement develops. It is interesting to note that
the amount of crying in infants from birth to 30 days of age has
been found to be related to environmental factors such as the daily
length of nursing, and the infants' living quarters, whether living
in a family home or a hospital nursery, indicating a reduction of
crying the more "supporting" the environment.

To the extent the crying-breathing integration breaks down
or never develops, that is to say, to the extent crying spells become repetitious, the infants' sucking problem will be overruled and
put away in the background, but all the same we will be confronted
with the infants' response to its ungratifying oral behavior. It
might be helped to integrate successfully its crying responses,
it might continue its crying spells although this line would probably
sooner or later be met by strong negative sanctions from its parents,
or it might start defending itself against its objectionable behavior
- through postural changes suppressing the crying response.

At this point we have to go one step further and ask what
the postural basis is for crying and how this specific basis may be
dissolved.

To answer this question we would have first to inquire
into which muscle groups are focal in the process of crying.

A most conspicuous gap still exists in our knowledge as
to the specific muscles involved in various basic responses. This

in spite of the groundwork layed down by Darwin in his The Expression of the Emotions in Man and Animals as long ago as 1873. His description of the facial behavior of infants during crying is summarized by Ames (1941) as follows:

"The infant when crying closes his eyes firmly so that the skin around them is wrinkled and the forehead contracted into a frown. The mouth is widely opened with the lips retracted in a peculiar manner, which causes the corners to be drawn down and the mouth to assume a squarish form, and the gums or teeth to be more or less exposed. The contraction of the orbicular muscles leads to the drawing up of the upper lip and the drawing upwards of the flesh of the upper part of the cheeks which runs from near the wings of the nostrils to the corners of the mouth and below them." (p. 239).

Ames (1941) maintains that since Darwin's publication relatively little study has been made of the crying behavior of infants, and that no report can be found in the literature which includes as detailed an account of general body activity in crying as Darwin gave of facial activity. To remedy this neglect Ames sets out to determine whether or not any specific motor behaviors, other than the traditional face-mouth behaviours, characterize crying in the human infant. Studying 13 infants at monthly age intervals during their first year of life, Ames emphasizes the involvement of the forehead, the cheek, the jaws, the eyes, the tongue, the nose and the extremities. And she concludes:

"Motor behaviors which commonly characterize crying in the human infant include body postures as well as traditional facial patterns described by Darwin. Crying is characterized by marked limb activity, greater leg than arm activity, unilateral rather than bilateral and flexor rather than extensor movements, and the breaking up of postures prevailing at the time of the onset. These behaviors clearly distinguish crying from noncrying. The marked flexor character of crying behavior may have a protective significance." (p. 247).

From the descriptions given it emerges that crying involves nearly the whole body. But which are the focal muscles involved? The immobilization of which muscles would enable the infant to suppress and subdue its crying behavior?

Somewhat different answers to these questions have been
offered by different observers. Reich (1942), for instance, in
pointing out that the repression of basic responses is never a
matter of individual muscles becoming spastic, but of muscle groups
forming a functional unit from an affect-expressive point of view,
maintains: "If an impulse to cry is to be suppressed, not only the
lower lip becomes tense, but also the whole musculature of the mouth,
the jaw and the throat; that is, all the muscles which as a functional
unit, becomes active in the process of crying." (p. 269). Perls et
al. (1951) emphasizes the contraction of the diaphragm as a mechanism
behind chronic inhibition of crying and sobbing, while Lowen (1958)
particularly stresses the elevation and the pulling backwards of the
shoulders, the tensing of the muscles at the root of the neck, the
longitudinal muscles of the back and the deep abdominal muscles, a
picture fairly similar to the one described by Ribble as characteristic of the negativistic syndrome in infants. Also the tonicity
of the musculature of the forehead, of the tongue, around the eyes,
and at the scalp at the top of the head, has been considered relevant
for the suppression of early oral responses.

Spitz (1951), in his treatment of psychogenic diseases in
infancy, subdivides these into two groups: psychotoxic diseases
and emotional deprivational diseases. The latter category he divides
further into the effect of total and partial deprivations, respectively. Stupor and coma, oral lethargy, diminished reflex excitability
throughout the body, is, according to Spitz, to be considered the
result of an open global and total rejection from the time of the
birth or possibly before. This condition will lead to death if no
treatment is given. Possibly, the condition is i dentical to marasmus
and infantile atrophy, diseases assumed to be responsible for nearly
half the infant mortality rate a few decades ago. The condition also
has some very striking similarities to a condition described by Spitz
(1945) as hospitalism. Here too, apathy, resignation, lack of
interest in external objects, lack of readiness to invest feelings,
and lack of protest to deprivation and disappointment, stand out as
central characteristics. On the other hand, the deterioration of
primary body reflexes seems to take place at a much slower rate,

parallel to the environment in the latter case certainly being depriving and impersonal, but not totally rejecting.

Generally, the younger the child, the more detrimental does deprivation of affective contact seem to be for its development, but during the whole of early childhood the separation of children and parents may have considerably injurious effects unless the child is given an opportunity to create emotional links with one or several other persons in its new environment.

We stated earlier in this chapter that contact between the infant and its environment will be effected particularly by its mouth. From the very beginning searching impulses seem to be mostly concentrated around the mouth, but after a while other parts of the body will be included. From the age of two months the infant will exhibit seeking and approaching impulses by means of its arms and eyes, and a couple of months later the beginnings of coordination of eyes, hands and mouth. The child smiles at faces, first on a reflexive basis, later as an expression of its search for contact and its dawning social orientation (Spitz, 1946).

The late oral phase of development.

At about the age of six months we glimpse new features in the child's behavior which makes it possible for us to say the child is about to enter a new phase of development. In making a division here we do not imply the appearance of totally new behavior patterns unconnected with earlier ones. As in other maturation processes the transition is gradual and continuous. Our reason for choosing the age of six months is partly the growth of the infant's first teeth, and partly its dawning ability to grasp different objects on a non-reflexive basis. Sucking is still a dominant element in its relationship to its environment, but its teeth and grasping ability imply that the external world can gradually be met in a new fashion. If we observe a child at this stage we discover that it likes to put things in its mouth; it looks as if it needs to bite and enjoys it; its hands reach out and take things, its eyes focus on and seem to "grasp" objects, and its ears capture aural impressions. The child is still receptive and responding, but its mode of receptivity and activity is changing.

Sucking gives place to chewing, to begin with perhaps mainly to offset the discomfort of teething, to relieve tension in the gums, but after a while as an independent source of pleasure and as a means of obtaining food.

As new oral activities become differentiated the infant's sucking impulses will spontaneously wane. Vocalizing and biting will appear as new avenues of pleasure-getting. The sight and the sound of the mother will replace the child's earlier satisfaction of tactual contacts with her. "When this stage has been reached with skilful handling and with a minimum of privation of the primary mothering", Ribble (1944) maintains, "the child gives up sucking activity spontaneously, and weaning is not necessary". On the other hand, if "children have been frustrated in their earlier sucking and in primary mothering, they (will) give up sucking later and less spontaneously" (p. 640).

As sucking gives way to chewing, we gradually also find another change; the swallowing whole of fully prepared food, to the biting off bits and working them over to ensure their assimilation. In short, the uncritically accepting of what is given gradually gives way to a more active, selective orientation toward what the environment offers.

The dawn of independence symbolized by the chewing is followed by an increasingly greater ability for specific mobilization of energy. This is especially manifested in situations in which the child's impulses are blocked or frustrated.

The growth of the teeth and the increasing control of the jaw muscles enable the child to bite off, to bite out and to bite at in the sense of snapping back. Here too, as in the early oral phase, we are probably confronted with a learning process, the integration of breathing and a new oral modality. Earlier the child had to learn to get and to cry and protest, now it has to learn to take and to aggress.

While frustration in the early oral phase led to diffuse struggle, rage, protests without any particular point of attack, the child will now be in a position to react aggressively in a more directed and focused manner.

As just indicated, a characteristic feature of what we call the late oral phase is the emergence of a <u>receptive self-providing modality</u>. But although the emergence of this modality is a part of the child's maturation process, it is not completely innate but dependent upon sufficient support being offered the child by its environment.

If the child's conditions for growth were inadequate in the early oral phase, this will strongly influence the child's ability to cope with the maturational changes taking place in this later phase. It has been pointed out that early oral conflicts may be reflected in the way the child is holding its body and in its general grasping ability. For instance, on the basis of extensive observations of institutionalized children, Fisher (1958) draws attention to the following pathognomic signs in the second half of the first year: "restless grasping", i. e. immediate release after prehension which does not allow systematic manipulation or combining of toys; "atypical grasping position", i. e. the upper arms lifted horizontally in abduction, bent at the elbows, so that the hands dangle passively near the ears; and, closely connected with these, unwillingness to assume and actually a resistance against, a sitting position, this in strong contrast to even 3 to 4-month-old infants who may often protest violently against a supine position, and become calm when they are held or supported in a sitting position and thus allowed a more extended and gratifying field of vision. Fisher emphasizes that these reactions are not to be considered as expressions of deficient physical maturity, but that they have an emotional basis which justifies the term "essential emotional immaturity reactions."

But also in the case where the child's early condition for growth has been satisfactory do the late oral phase present challenges which have to be met in their own right.

For instance, weaning may easily give rise to numerous critical situations. If it does not proceed flexibly in accordance with the child's growth and development, if towards the end sucking occurs in an atmosphere of impatience and tension, biting impulses will easily be aroused. The child's mother, perhaps tense and un- certain before, will react by even greater uneasiness which in turn

may be transmitted to the child. If the mother withdraws when the
child bites or tries to bite, this may often mean much more to the
child than a treatened loss of food. Its whole existence may be
considered in danger.

The growth of the teeth and the increasing control of the
jaw muscles enable the child to bite. This newly acquired ability
can easily become a two-edged sword as far as the child is concerned.
Aggressive biting in small children is often regarded as extremely
improper and unseemly behavior. The child's impulse to bite in
response to frustrations may easily become a boomerang which strikes
back in the form of new threats and new frustrations.

A mutual regulation of behavior between mother and child
at this stage implies not only that the child's biting is not
crushed, that the child's weaning is not too sudden and rough, but
also that the child's dawning impulses towards taking and self-
provision are supported and encouraged. An overprotective, symbiotic
attitude on the part of the mother may have a suffocating effect on
the child's growing self-assertion, and may in fact probably be just
as impulse blocking as a direct rejection and withdrawal.

As alluded to earlier, moving into the chewing stage the
child's ability to release excitement increases, although it is
likely that the coordination between its newly emerging tension
relieving activities (chewing and biting) and its breathing, initially, is of a rather brittle nature. In fact, we assume that the
child has to learn to integrate biting and breathing for its biting
to develop into a tension relieving aggressive response.

A fragmentary biting response, a response unintegrated with
breathing, will most likely leave the child close to a "trigger state"
so that the response will repeat itself over and over again.
A repetitious biting response will at least in our culture, commonly
be met, if not by overt rejection and punishment, by appeals mixed
with threats of loss of love or love object. Thus, the child's
emerging biting impulses will easily result in experiences furthering
the development of an ambivalence on the part of the child, an ambivalence between impulses to bite at and to hold on to, the latter
element resulting from the child's growing attachment to its parental

objects. And as this ambivalence increases in strength, the child
is forced into a defensive position as regards its objectionable
responses. Again, we are faced with postural modifications dissolving the basis of the responses in question.

The most narrow and active postural defense against biting
impulses is a static clenching of the jaws. By tensing the jaw
muscles the child confines his oral aggression by locking its avenue
of expression. However, by so doing he also directs it toward
himself. Instead of aggression and taking, the child's mode of
activity changes into one of keeping in possession, keeping hold on
and hanging on to. That is to say, the child's self-providing
tendencies get overruled by a clinging, adhesive mode of behavior.

As in the early oral phase, we may distinguish between
different forms of postural defenses.

Discussing postural defenses to biting and chewing, Perls
et al. (1951) emphasize first and foremost the "hanging on bite",
but they add that the "hanging on bite" not only will involve the
jaws, but also the throat, the chest and the eyes, bringing about
a fixed stare, preventing a 'piercing' glance.

Also regressive tendencies may come to the fore, tendencies
toward sucking and licking. In some cases the underlying ambivalence
may give the clinging a distinctly sadistic tone, in other cases,
where the child's clinging doesn't produce relief, where it is faced
with a situation where its significant object has completely withdrawn and in which there is no possibility of making emotional contact
with another object, the child may move into a state of grief and
fretful whimpering, pass over into a period of contact-refusal,
insomnia, developmental arrest, generalized suppression of activity,
and still later relapse into a state of lethargy, apathy, developmental retardation and melancholy. In this latter psychotic-like
phase, the child's face becomes expressionless, distant, and its whole
orientation depressive and withdrawn. Spitz (19;7) has described
this developmental sequence in great detail and proposed the designa-
tion anaclitic depression. The deterioration process taking place
parallels, to a certain extent, the prenatal regression found under
severe deprivational conditions during the early oral phase. The

sequence seems to be alike, but the rate of speed and the deepness
of regression is perhaps somewhat greater in the first instance.

Far from all deprived children in the latter half of the
first year develop the syndrome described by Spitz as _anaclitic
depression_. In fact, Spitz found children with a pretty good early
object relationship more susceptible to this reaction pattern than
children with a less satisfactory object relationship. Furthermore,
not all deprived children in the latter half of the first year
experience the severity of deprivation implied in a total object
loss.

Braatöy (1954) in a case study of a thirty-three-year old
woman, alludes to a probably more frequent form of regression, the
substituting of biting and self-assertion with sucking and self-
consoling oral activity. He states:

> "In a period of her analysis, every deep inspiration was
> suddenly interrupted by a drawing in of her abdominal wall.
> It presented itself as a semiorgastic reaction but was not
> released until I discovered that this respiratory inter-
> ruption was more related to a sucking inspiration. The most
> inaccessible part of this reaction was the position of the
> tongue. This part (the sucking of the tongue close to the
> roof of the mouth) was completely unconscious to the patient,
> probably because it was of such long standing and was performed
> automatically. The important function of such details... is
> that complete emotional release sometimes is not achieved before
> such fragments are mobilized." (p, 181).

We don't want to pursue any further the postural basis for
the inhibition of biting and self-providing impulses. In fact, our
knowledge in this area is still very limited and fragmentary.

To recapitulate, as we see it, the impulses being crystal-
lized in the late oral phase are intrinsically neither destructive
nor sadistic. The child's activity may, however, have destructive
consequences. The child has not yet learned to distinguish between
objects d'art and objects with which it is permitted to play, between what it may throw about and bite and what it must leave alone.
A much debated question is whether parents ought to protect their
children from experiencing anger. According to the view presented
here, the mobilization of aggression may, under certain conditions

have a beneficial effect on the child's development, as long as its
spontaneous anger is not rejected or condemned, and its anger is not
of such a character that it prevents or disrupts an integration with
breathing.

The child's dawning selv-providing orientation is expressed
in several areas. It will spontaneously turn from sucking to chewing, from being fed to feeding itself, from receiving things to
grasping them itself. It will reach out towards its environments
more specifically than before, with growing purpose and with greater
confidence in its own powers.

The early anal phase of development.

Normally we can observe the child's first grasping movements
before the age of six months, but not until it is nearly a year old
is it able to coordinate its thumb and fingers so that it can hold
the objects it grasps. The one-year-old is able to grasp and pick
up objects, inspect them and let them go. The components in this
interaction between "taking" and "letting go" seem to develop more or
less simultaneously. While the former impulse pattern belongs, in
our opinion, to the late oral phase, we are inclined to ascribe the
latter to the early anal. These two phases seem to be more closely
related in time than any others (Freud, 1951). So much so that a
defensive postural solution of late oral conflicts probably to a
large extent will be patterned by the way the child learns to cope
with its incipient anal impulses.

During the first months of a child's life fairly undifferentiated mass activity is characteristic so far as the child's
defecation process is concerned. The infant reacts as a unit, and
all kinds of strains immediately affect its appetite and gradually
also its digestion and bowel movements. This displacement of sensi-
bility downwards is the result of a general cephalocaudal process
of maturation.

During the first months of a child's life the intestine
is emptied more frequently than it is later on, and the excrement is
quite loose. However, at about six months a change starts to take
place. The fluid in the upper part of the large intestine is more

effectively absorbed, and movements in the lower part of the
intestines which divide the contents up in sausage-like lumps
begin to occur at regular intervals.

Defecation thus acquires a certain regularity as well
as form and consistency. When the contents of the intestine
enter the rectum, pressure arises which automatically releases
a coordinated contraction of the abdominal muscles and relaxation
of the muscles closing the rectum so that the lumps are expelled.

The movements in the large intestine attract the child's
attention toward the process of excretion. The mouth still
represents an important source of pleasure, but gradually the anus
acquires great significance. The movements in the large intestine
seem to be experienced as troublesome, on a par with hunger, i.e.
pain in the stomach, and elimination as enjoyable, relaxing and
pleasant, like eating. Just as an infant's receptivity starts out
from a congenital searching and sucking reflex, so is elimination
initially a congenital excretory reflex. In contrast to the former,
this latter reflex does not function completely before the child is
several months old. But in agreement with the former, here too, we
are probably faced with a learning process, the child's having to
learn to coordinate his breathing and defecation in order for the
latter to be fully enjoyable, relaxing, and relieving. In fact, we
may think of this learning process as fundamental for the child's
ability to assimilate, integrate and experience the letting go
modality as part of its personality.[1]

An infant does not have any clear consciousness of the
surface of its own body. Its ability to distinquish between internal and external stimuli is rudimentary. While its first months

---

[1] Basically the learning process in question may be considered
as reaching into the very process of defecation (and micturition)
- the child having to learn to expire without "expiring", i.e.
to keep the glottis closed and diaphragm contracted while
contracting the abdominal muscles so that the pressure increases
within the abdominal cavity is not transmitted upward but
brought to bear on the intestines and pelvic organs.

of life are characterized by a complete psychosomatic symbiosis
with the mothering object, it will gradually be able to make
distinctions between what is internal and what is external, and
to discover itself as a separate object. The excretory process
seems to play an important role in this connection.

The excretory process stands for more than a passive
experience of pleasure. It seems to the child to represent a new
area of contact with the environment. Its excretory products
awaken the child's positive interest where taste, smell and touch
are concerned, and the process in its broadest sense seems to be
an important testing ground for the child's dawning self-discovering
tendencies. According to some scholars, by letting go of the anal
product, which is part of itself, the child learns to let go, or
to untie, its symbiotic relationship to its mother. Our own conceptions coincide very much with such a view.

As just indicated, we conceive of the early anal phase as
the period in which <u>an eliminative selv-discovering modality</u> emerges
in the child, granted it is offered sufficient emotional support by
its environment. That is, emotional support is needed to no lesser
extent than in previous phases. Lack of consideration for the child's
rhythm, lack of understanding of its feeling of belonging to its own
product, a toilet training begun before the child has reached a
sufficient level of maturity, the meeting of its dawning interest
in its excretory product by disgust, disapproval and punishment, may
easily give grounds for conflicts, conflicts impeding or blocking
the learning process confronting the child.

Insufficient support may take the form of an overt denial
of love, of an over-protective, indulgent cleanliness training, in
which the child is denied all possibilities of experiencing pleasure
and experimental action in the direction of "letting go." Being
interfered with, the child's defecation may lose its initial rhythm-
icity, become irregular, periods of constipation being followed by
periods in which the time interval between defecation sharply
decreases.

As the child grows older, its eliminative impulses will,
when the child is confronted with frustrations, acquire an aggres-

sive-expulsive aspect very much similar to the changes taking place
on the oral level. Confronted with frustrations, the child will
react with rage and protest by kicking, trampling, slapping, and
knocking away. And as was the case with aggressive biting, the
anal aggressive-expulsive tendencies too will have to be integrated
with breathing in order to be really tension relieving.

Not being able to attain such an integration, the child's
"aggressive" self-assertive behavior may acquire a repetitious
character, a repetitiousness that may change the parent-child relationship into a battlefield and the child's excrements into ammuni-
tion. Being confronted with increasing threats of punishment, threats
of loss of love and loss of love object, the child will be caught in
a growing ambivalence between simultaneous impulses toward expulsion
and retention, between letting go and holding back, between aggression against and attachment to. And again various postural defenses
are open to the child, one postural solution going in a hypertonic,
another in a hypotonic direction.

Peculiar difficulties regarding the letting go modality
have been observed in children between 12 and 18 months of age.
Hoarding in these children have become an aim in itself, and their
reaction to demands to give up toys or other objects is most remarkable, even giving up toys in a purely indirect way, such as when
piecing together a puzzle or putting bricks into a box, brings about
violent reactions, the children often throwing the toys away in
anger.

The intimate connection between postural factors and bowel
habits has been described by Gesell et. al. (1943). Commenting upon
the same question, i.e. the postural aspect of anal function,
Braatöy (1954) states:

"The trouble with too early toilet training is that it tries
to teach the child differential relation to sphincter mechanisms
at a time when spincter relaxation cannot be achieved without
general relaxation of the lower limbs. To relax on the pot
includes at this age the risk of falling off the pot. The child's
stubbornness is, in such a situation, aided by respectable motives"
(p. 173).

Specifically, according to Braatöy, anal problems are connected with motor control or lack of control of the legs. This last point is emphasized by Lowen (1958) too. According to Lowen an enduring result of premature parental insistance on excremental cleanliness is to force the child to employ the buttocks and the thighs (the levator ani, the gluteals and the hamstrings muscles) to gain anal control since the sphincters have not yet fully developed. And Lowen suggests that it is these very muscles which get involved when the child for one reason or another is forced into a defensive-repressive position toward its anal-expulsive impulses.

Very much the same position is expressed by Reich (1949), stating that "The muscular holding back of the feces is the proto-type of repression in general and is its initial step in the anal sphere." Reich too, mentions specifically the musculature of the buttocks but doesn't exclude that a spasm of the anal spincter too may emerge.

In his description of the postural inhibition of anal expulsive impulses Lowen not only points to the mechanism of the child squeezing the buttocks together, but suggests that this very mechanism is only a fragment of a broader postural defensive pattern including the tensing of the extensor muscles of the thighs and calfs, tensing of the throat, the abdominal muscles, the muscles of the lower lumbar region, pushing the pelvis forward and upward, and tightening the shoulders.

Granted the child is forced into a regressive solution, we may conceive of the child's anal skeletal muscles getting a flabbish quality, and the child's orientation becoming increasingly oral and symbiotic.

This latter conception corresponds to Spitz's (1949) de-scription of children who after having experienced a comparatively satisfactory affective relationship during early infancy, later on met severe deprivations in the anal phase. In these children, Spitz maintains, not only tendencies to coprophagia were pronounced, but also an inwardly directed "abstract" self-devotion, often mixed with a suspicious, almost paranoid attitude.

To the extent a child is caught in an unbearable
ambivalence between letting go and holding back on the oral as
well as on the anal level, and reacts to the ambivalence with a
postural immobilization of all out-reaching aggressive impulses,
we are, according to some observers, confronted with the structural
basis of the sadomasochistic personality formation. The blocking
and repression of aggression will, it is maintained, establish a
self-perpetuating drive system. The break through, or perhaps more
correctly - the pushing through - of aggressive impulses on the
oral level will give rise to biting reactions in the form of crushing
and biting-holding-on-to, and to the emergence of fantasies in dreams
and daydreams of a cannibalistic nature; and the breaking through on
the anal level, to smashing and squeezing, and to the emergence of
fantasies concerning destructive explosions. These categories of
behavior will also be pronounced in the perceptual realm.

The masochistic counterpart to the above picture would
consist of the aggressive element being projected outward and turned
inward, the repressive forces being too heavy to allow for any
discharge of self-represented aggression either in fantasy or reality.
Closing itself up for self-assertive impulses, confronted with
continuously active longings for contact, but unable itself to reach
out, the child is left with seeking relief in submission and servil-
ity, in imploring approaches of others by means of whining and complaints, and by fantasying itself into a suffering or humiliating
situation.

There is much evidence that early anal training impedes
rather than stimulates a child's development in the direction of
excretory control. Moving into the question of control, we think it
is most important to distinguish between two different types of
control: a mechanical one and a more functional, flexible one
anchored in an inner autonomy. The latter form of control presupposes that the child experiences his excretory process as an integral
part of himself, but also that he has achieved an active orientation
toward this very process. The emergence of this latter orientation
we will consider a central aspect of the late anal phase.

The late anal phase of development.

We stated earlier that the one-year-old already has a certain ability to hold on to and let go of various objects, but that the interplay between these modes is not nearly well enough developed for us to speak of real will-power or self-control on the part of the child. In the course of its second year of life, tremendous development takes place in precisely this area. Several features of the child's general behavior indicate this development. The child shows increasing interest in playing with objects, in taking toys to pieces and putting them together again, in filling up and emptying, in building up and pulling down. It is no longer elimination which gives satisfaction so much as the pleasure of shaping "elimination" to a product. Pride in its own accomplishment enters into the child's life. The child no longer concentrates on taking and letting go, but on how, where and when. The main pleasure is in shaping and forming. A precondition for this pleasure probably is the feeling of having something to make, of having some-thing to let go, but which can also be kept, formed and reformed be-fore the letting go.

Parallel to this growing capacity for functional producti-vity, i.e., productivity for its own sake, we observe a rapid devel-opment in the child's will power. A very characteristic feature becomes the child's rejection of help from others. Halfway through its second year it begins to express negations. Its incipient ability to say no is increasingly typical of its behavior. The negation seems to give the child pleasure and to imply its growing consciousness of itself as an independent entity, and its behavior, the testing out of a new behavioral dimension in its life. In fact, Spitz (1957) considers the child's no its first really abstract conception, the very beginning of its abstract thinking, and that the child's later ability to say yes basically hinges upon its also being able to say no.

Keeping to our earlier train of reasoning we will assume that the child's ability to say no with full orchestration and self-representation again implies a learning process, an integration

between breathing and bodily expressions. And furthermore, that the
child's inner autonomy, self-control, and productivity is decisively
shaped through this learning process.

We find perhaps the first impulses in the direction of
forming and shaping in connection with the mass movements of the
large intestine, and later too a link exists between excretion and
forming. A child's excretory product is its first form of productivity. The defecation process itself provides a central testing
ground for its self-control. The excretory product awakens its
interest, first as something to let go, and latter as something it
can manipulate and about which it can make decisions. Its manipulation and decisions imply that an inner autonomy is beginning to take
form. It is still highly dependent upon its immediate environment.
In fact, children between the age of 18 and 24 months of age often
react the most severely to separation from their parents. It is as
if the child's incipient independence and will-power make it particularly dependent upon continuous support from its parents.

From what has just been said, we conceive of the late anal
phase as the period in which an eliminative self-deciding modality
emerges, granted the child is offered sufficient emotional support by
its environment.

However, as prevously, the support needed is not always
present. The possible area of conflict between child and environment is large. The child's play and its desire for activity may
clash with its parents' demands for cleanliness, duty and tidiness;
its wish to decide for itself may meet with lack of understanding
and the opinion that the child is obstinate; its pride in its own
products may meet with laughter and indulgent irony, etc. And as
previously, not being met with sufficient support from the environment the child's newly emerging impulses may become repetitious in
character. Its repetitious no will take the form of stubborness and
persistent negativism.

Because of the child's increased muscular coordination its
protest and self-assertion will be more violent than previously.
Its anger will be expressed in screaming, banging, stamping, throwing

things about and kicking them in open defiance. Of course, these
responses, and especially if they get repetitious, may have such
a challenging effect on the parents that great efforts will be
invested in crushing the child's defiance, crushing it by caning,
threats to send the child away or deprive it of love and affection.
What is at stake is the child's inner autonomy, its spontaneity,
and its incipient control over its own functions. Although control
of anal matters certainly is an important aspect, it is probably
only a fragment of the total issue - the child's general psychic and
motoric control.

If the child is rejected or threatened sufficiently severely
it will as before start protecting itself against its own impulses.
The conflict between the child and its environment will be internal-
ized as an ambivalence within the child. The ambivalence in question
will be expressed as simultaneous impulses to defy and obey, to aggress
and to hold back. And, as previously, if the conflict gets too overwhelming, a solution may be sought in the immobilization of the
postural basis for the child's objectionable responses. Again we are
confronted with the question as to what represents this postural
basis?

Braatöy (1954) suggests that since the child and the problem
of control confronting the child is a general one, involving his whole
body, a postural defensive solution will also have to be of a general
nature involving nearly all skeletal muscles. Reich (1942) indicates
that the solution sought being the holding back of impulses, the back
muscles, the musculature of the neck, and between the shoulder blades,
will be particularily involved. Perhaps the most extensive discussion
of the problem is offered by Lowen (1958). Lowen indirectly suggests
that a precondition for the child's no is that it has arrived at a
stage in which it can assume a standing up against erect posture, and
that the no represents an incipient ability to shut itself off from
frustrations - in other words, that it represents the child's initial
ability to meet frustrations not only by fight or flight but by
controlling, postponing, and modifying its spontaneous affective
responses. The process in question, Lowen maintains, consists of
the child being able to armor itself - psychologically and somatic-

ally somatically by tensing the muscles at the front of the body, the chest and the abdominal walls. By so doing, the child may obtain situational protection and control. However, what happens when the child is forced into a defensive solution is that the situationally induced armor becomes permanent, the child locking itself into a posture consisting not only of the tensing of the muscles at the front of the body, but the tensing of these muscles being followed by a simultaneous pulling back of the shoulders and the pelvis, putting the front muscles on stretch in a vise-like fashion. By stressing the importance of the shoulder muscles Lowen's view coincides with Reich's conception, and by stressing the global involvement of hypertensions, agreement is also present with Braatöy's formulation.

Looked at from the point of view of respiration, the viewpoint also makes sense. As pointed out previously, first when the child assumes a firm erect posture do we find an increasing downward slope of the ribs, accompanied by a downward sinking of the abdominal organs, and first after this point has been reached, around the age of two or three, do we find the gradual strengthening of the thoracic muscles making them susceptible to more forceful movements, and probably also, to immobilization of movement. This latter case, as alluded to previously, most likely implies a respiratory learning process. A process which, if not interfered with, involves a learning to integrate breathing and shifts in the tonicity of the thoracic musculature, to push the breathing down in the abdominal region exclusively.

By armoring itself permanently, the child turns to a mechanistic type of control. It loses its intimate relationship to its own productivity and creativity. Its pleasure in the process of shaping and forming easily becomes blocked in favor of an expedient and ambitious attitude toward achievement, its self-esteem limited to its productivity, or its productivity perfectionisticly oriented with a preoccupation on shaping at the expense of making and creating. The "back-ache syndrome", as referred to by Holmes and Wolff (1950), has some of these very qualities.

As in earlier phases, so also in the late anal phase, we may conceive of a sort of regressive hypotonic postural solution, the child abandoning its "standing up against" attitude and posture, and withdrawing into a "don't care" attitude towards forming and shaping, i.e. into a careless, impulsive, "spineless", antiperfectionistic "letting go" mode of behavior. Posturally speaking, we would in this instance expect the child's chest to be soft and mobile, its buttocks and thigh muscles to be rather hypotonic, and far from showing marked stiffness and rigidity in the back and in joints of the knee and the ankle, most likely they would be easy to bend. We are here reminded of Deutsch's description - previously referred to - of the postural picture characterizing what he calls anal eroticism.

We stated earlier that in its second year of life the child's self-control and autonomy go through a period of fast development. We find many characteristic features of this development in its third and fourth years as well. Language begins to play a greater part in the child's orientation toward itself and its environment. Its behavior will gradually become more intentional. It begins to formulate in words and sentences what it wants and does not want, and to experiment in roleplaying as to what it is and is not. This again is a kind of shaping, but considerably more differentiated than before. The late anal phase lasts for a comparatively long time and includes a large number of specific activities. A common feature of them all, in our opinion, is the child's testing of its powers of decision and its pleasure in the process of forming.

The early genital phase of development.

The age of three is a violent period of expansion as regards a child's experience of its own will-power. The age of four is also a period of expansion, but in a slightly different area. Exploration of the external world takes place more intensely than before. The child's mobility increases considerably. It can walk and move without concentration. A child is normally able to walk when it is a little more than a year old, but not until now does walking become fully automatic. Further development on the purely mental plane takes place parallel to growing motoric mastery: the child acquires capacities for symbolic activity and the use of speech. It begins to ask questions and to take an interest in cause and effect. By means of locomotion and speech it attempts to widen its horizon and increase its knowledge and understanding. This period is in many ways a daring voyage of exploration.

These tendencies towards expansion, exploration and mastery are also expressed in the child's imitation of other children and adults. Its incipient role discrimination is followed by interest in the difference between boys and girls. The sexual organs attract attention as a source of pleasure. Masturbation may have begun at the age of two or three. But now the child's interest in other children gradually develops, and sexual play among children is common at the age of five. The child's experimentation with itself and others gives it an increasing awareness of its place in a wider social context. Its self-awareness acquires a new quality and its initiative a new dimension.

The above remarks apply to both sexes. The tendency in the direction of making, of exploration and experimentation is the same, but the mode of behavior gradually becomes somewhat different. A crystallization of sex-specific impulse patterns seems to take place at about the same time as the initial perception of anatomical sex differences. According to some observers, in boys of this age a change will often occur in their attitude to the opposite sex, a development of a more indulgent, protective and chivalrous type of behavior.

Sexual play is mainly concentrated on mutual feeling and touch, essentially a form of bodily play. However, among children of five and six copulation-like play patterns are also observable, patterns strongly reminiscent of adults sexual behavior. Since such patterns have been found to occur also in environments where the children probably have had no opportunity to observe adults sexual behavior, it is likely that we are confronted with responses which have inborn, instinctual roots; responses in boys of an

"introducive" and in girls of an "embracing-enclosing" character.
We are going to consider these responses as focal ones for the early
genital phase, that is to say, we believe that granted sufficient
emotional support from the environment the early genital phase in
boys is characterized by the emergence of an introducive self-
expanding modality, and in girls, by the emergence of an embracing-
enclosing self-expanding modality.

Following our earlier train of thought we believe that in
connection with genital responses too we are confronted with a
learning process, with a learning process involving the integration
of the responses in question and respiration. As previously, we
think that the emotional support offered the child from its environment is of great importance for this learning process, and above all,
the child's opportunities for social and bodily interplay with children
on its own age level. It has been suggested that childhood peer-
interaction and bodily play in monkeys represent a prerequisite for
their later smooth sexual functioning. Probably a similar connection
exists in humans too.

As previously, a number of potential areas of conflicts
exist between the child and its environment.

Given limited opportunities for sexual play with other
children, or if such play is punished, rejected or condemned, the
child's sexual interests will become one-sidedly focused on its own
parents. We will be faced with an oedipal situation. How such a
situation is solved varies, of course, considerably. The child's
genital impulses may be accepted, their expression toward its peers
encouraged, or they may be laughed at, condemned and rejected.
Parental rejections may take many forms: it may consist of overt
punishment, or seductive tenderness and bodily stimulation, followed
by abrupt rejection of the child's active responding impulses, etc.

We would expect that the child's genital responses, not
being integrated, will be "triggered off" again and again. However,
being deprived of appropriate environmental conditions, the child's
genital responses will probably not come to the fore directly but in
indirect ways. In spite of the child's greater tolerance of frustrations, its greater ability to postpone immediate expression of its

impulses, it is still largely dependent upon its environment.
In the case of boys the emerging genital impulses will be linked
to an ambivalence, an ambivalence between an intrusive self-
assertiveness and tendencies of holding on and holding back, and
in girls, an ambivalence between closing in, i.e. moving forward-
embracing, and closing up - holding on to. And as previously,
given the development of a sufficiently strong internal conflict
in the child, we would expect it to seek protection against
repetitious or continuous frustrations in postural modifications,
in modifications dissolving the reaction basis of its frustrating
responses.

Again we are faced with the question of what constitutes
this reaction basis. One fragment of course is the tensional state
of the genital musculature itself - of the bulbo-cavernosus and the
ischio-cavernosus muscles, - but intimately involved are also
(most probably) the bodily parts to which the genital organs are
connected, specifically the abdomen, the pelvis, the legs and the
spine. Earlier when we used such terms as closing in, holding back,
embracing, etc., we were referring to general behavioral modalities,
but also to what we assume to be specific motility-modes of the pelvic
region of the body.

An immobilization of genital responses can take different
forms. In both boys and girls we may distinguish between an active
repressive and a more passive regressive solution.

Discussing the somatic location of genital inhibitions,
Reich (1942) states:

"... one usually finds that the pelvis is fixed in a retracted
position. Sometimes, an arching of the spinal column goes with
this reaction, the abdomen being pushed out. A hand can be
placed easily between the lower back and the crack. The
immobility of the pelvis gives the impression of deadness...
The patients are unable to move the pelvis. If they try to
they move abdomen, pelvis and thighs in one piece... As a
rule, they fight intensely against moving the pelvis by itself,
particularily against moving it forward and upward... Such
patients always suffer from a severe disturbance of the sexual
act. The women lie motionless, or else try to overcome the
blocking of their vegetative mobility by forced movements of
trunk and pelvis together. In men, the same disturbance takes
the form of quick, hasty and voluntary movements of all of the
lower body." (p. 304).

In addition to or supplimentary to the retracted position
of the pelvis, Reich pays specific attention to the downward
fixation of the diaphragm, the contraction of the musculature of
the abdominal wall, the pulling up of the pelvic floor, and the
tensing of the superficial and deep adductors of the thighs,
resulting in a pressing together of the legs.

The close connection between leg postures and genital
defenses was alluded to earlier in our discussion of Deutsch's
psychoanalytic observations. The same connection has been commented
upon by Braatöy (1954) as follows:

"... the crossing of the legs on the couch gives the analyst
information in the very first hour. In women it is practically always a defensive attitude with a sexual nucleus...
in some situations it may be completely reasonable for a
woman to be on the defensive in sexual matters, but she may be
on the defensive also in situations where this is not necessary.
In that case she uses energy and strength without reason. This
may be demonstrated by a posture kept up everywhere and always
...the habitual posture may demonstrate a tension which acts as
a suppressive force all the time."

Lowen too (1959) time and again mentions the same relationship and asserts that genital defenses in boys can take the form
either of a phallic-narcissistic or a passive-feminine solution;
in girls, the form of either a masculine-aggressive or a feminine-
hysterical solution.

Characteristic of a phallic-narcissistic solution, Lowen
maintains, is the omnipresent determination, the "willed" quality
of all feelings, lack of spontaneity, self-imposed responsibility,
compensatory self-confidence and aggressive courage, persistence
and insensitivity; and on the somatic level - rigid retraction of
the pelvis; rather tight legs; narrowing and contraction of the
hips; braced, raised and squared off shoulders; a stiffening of the
back and the front of the chest; and a set, hard jaw. Dynamically
speaking, a phallic-narcissistic solution means that genital introducive impulses, i.e. impulses toward leading into, are modified
into the form of forcing into. Instead of what we may call an ac-
commodated intrusiveness, we find urges to push oneself in indis-
criminately, also where not invited or desired.

Characteristic of a passive-feminine solution, Lowen maintains, is a profound eagerness to please, lack of decisiveness and strength, exaggerated politeness and compliance, unwillingness to take responsibility or initiative; and on the somatic level — a gentle and humble appearance, the surface muscles of the chest, the abdomen and the legs being relatively soft and hypotonic, the deeper muscles — specifically of the pelvis, being tense, the shoulders being narrow, the jaw withdrawn, and the eyes and the face without strength and vigor. Dynamically, the passive feminine solution implies an avoidance of masculine assertiveness, both of an intrusive as well as of an introducive nature. Consequently, a passive feminine man cannot represent his masculinity directly, either he would have to function as a father to a dependent woman or as an infant to an older and stronger female.

Turning to the masculine-aggressive pattern in girls — Lowen states as characteristics, competitiveness, overt aggressiveness, penis envy, avoidance of tender feelings, and a compensatory self-sufficiency; and on the somatic level — deep seated spastic conditions in the pelvic region, rigidity of the thighs and the legs, the shoulders and the pelvis being narrow and pulled forward, and the thorax showing severe rigidity. In his comments on the dynamic basis of the masculine-aggressive pattern, Lowen emphasizes the greater vulnerability confronting girls as compared to boys due to anatomical conditions, the perception of the vagina being blurred by an initial perception of something lacking. Gradually the normal female becomes aware that she is a girl and not that she is not a boy. If she meets with a parental attitude rejecting her femininity, her vagina may be repressed as a perceptual entity, and her genital sensations focused on the vulva and the clitoris. Given such a basis, the clitoris may take on a phallic quality, with "penis envy" and "inferiority feelings" emerging as strong components of the personality. The masculine-aggressive pattern is considered the outcome of such a process turned in a compensatory direction.

By a feminine-hysterical solution Lowen refers to a psychological pattern in which passivity, lack of deep affect investment, wishes to be loved and protected, are prominent.

Often a pattern of sleeping-beauty-like, angel-like, or nice-little-girl-like behavior will be pronounced with emphasis on denial, innocence and pollyanishness. Somatically, characteristic features are the immobility of the shoulders, the shoulders being high and narrow, the retraction and the tightly holding of the pelvis, the presence of tension in the vaginal muscles and in the adductors of the thighs and in the legs generally. Commenting upon the dynamics involved in the structure, Lowen points out that the passivity found has an aggressive overtone, by invoking protection and approaches, it satisfies narcissistic wishes of getting without giving, of teasing others to believe they get when they really don't. Developmentally, the pattern stems from the oedipal situation, Lowen maintains, from a situation in which the girl's sexual impulses toward the father, after being rejected, become suppressed and repressed, and so also the anger aroused as a result of the rejection. In short, a pattern emerges in which sexual desires and longings become blocked by disappointment and anger, and the anger and disappointment, blocked by longings and desires. Both aspects of the conflict being repressed, it leaves in its wake a permanent ambivalent attitude towards males. By taking a sexual submissive position later on, the female may reestablish the oedipal relation in a way satisfactory to her deepseated infantile fantasies of revenge. In other words, the ambivalence becomes expressed in the formula: "Please take initiative, but I don't want to give you anything." By inspiring expectations which are subsequently disappointed, indirect satisfaction is obtained.

In the preceding paragraphs we have partly anticipated the long term consequences of certain early genital conflicts. Turning back to the early genital phase proper, we remarked earlier that this phase, in both boys and girls, represents a violent period of expansion with regard to physical and symbolic activity. Its lasting effects probably not only involve sexual attitudes in a narrow sense, but also the basis for the boys' and girls' later ability to manifest courage and initiative.

The child's imagination at the age of five or six will often make it at cross purposes with reality. Its lack of ability

to discriminate between fantasy and reality is manifest in its reactions to external obstacles. At the same time the imitation of
adults occupies a central position in the child's imagination.
Conflict with its environment may induce defensive identification
with the threatening object. Defence patterns emerging in this
period will probably be closely linked with identification processes
generally.

An important feature of the early genital phase is the
child's interest in and orientation towards organized play and the
company of other children. Its incipient independence is expressed
in an increasing preference for playmates of the same age rather than
for the company of sibling and parents. This is particularly notice-
able from the age of six. Some observations made in England during
the last war illustrate this point most strikingly. During the
evacuation from London it turned out that children of primary school
age showed fewer adjustment problems and behavior difficulties if
they were evacuated with friends and peer groups than if they were
evacuated with their families. The case was the reverse with
younger children. We can therefore say that the age of six is a
milestone in a child's life. Granted it has been sufficiently supported during the preceeding years, it now seems to have reached a
stage where it can not only tolerate separation from its parents,
but can also face loss of contact and rejection with considerably
greater flexibility and without being forced in conflict situations
into a permanently defensive position. The child has not yet developed
complete independence and social identity. Final maturity in these
areas does not occur, in our opinion, until puberty, during the socalled late genital phase, although we don't want to underate the
importance of the ground work being layed down in the interval between
the end of the early genital phase around the age of six, and the
beginning of puberty around the age of twelve.

All through childhood we are probably faced with synthesizing processes, cohesive forces of a maturational nature acting
on the postural solutions arrived at at earlier developmental levels.

Some remarks in summary and conclusion.

The foregoing descriptions of the child's various phases of development parallels Erikson's presentation on some points, while on others we have relied more on classical psychoanalytic conceptions. Our description differs from Erikson's on the following points:

In the first place, and this is mainly a terminological difference, instead of talking about organ modes, we have chosen the term phase modalities, by which we mean a generalized mode of reacting, i.e. a pattern of going at things, seeking relationships, stimulation and contact with the environment, stemming from innate properties of the child's organismic development.

We fully agree with Erikson in the necessity of taking cultural and social-environmental factors into account, but we nevertheless feel that Erikson in his deviation from the orthodox psychoanalytic model has moved a little too far into a noninstinctual position, while at the same time basing his conception on primary drive systems. Basing our view on observations that babies usually show greatest attentiveness (susceptibility to external stimulation) when inactive and awake, and on Irwin's findings that babies are more frequently awake and inactive 15 minutes after being suckled than before; on Spitz's studies on the smiling response; Ribble's observation of spontaneous weaning in children given opportunities of self-demanded feeding; observations from other cultures that toilet training does take place in spite of parental ignorance; Anna Freud's observation of spontaneous copulation-play among children in a residential setting, and observations of a number of children being "self-regulated" by their parents, we think it is reasonable to assume a somewhat stronger maturational foundation of "competence motivation" than Erikson suggests.

In another way we consider ourselves somewhat less orthodox psychoanalytically in orientation than Erikson: In contrast to both models, we think that the ego groundwork for getting to be the giver in social interaction is to a certain extent present in the baby at birth, that a supporting environment can save the baby from a traumatic change in connection with the eruption of the teeth, and also from traumatic experiences in the anal and early genital phases of development. In our presentation we have emphasized consistently the reaching out, contact-seeking aspect of various primary impulses, granted sufficient support is offered by the environment.

In accordance with the orthodox model we have distinguished between five pregenital phases, each characterized by pleasure in a certain mode of activity, although the modes of activity we consider "basic" for the various phases differ to a certain extent from the orthodox model (see next page).

Figure III gives a schematic survey of our thinking in this respect.

As previously mentioned, Erikson strongly emphasizes the many possibilities for conflicts emerging between the child and its environment, and the subsequent possibility of an inappropriate mode growing into dominance and influencing the child's psychic functioning and its ego development. The growth into dominance of such an inappropriate mode parallels what we have designated a defense solution.

We do not want to reject Erikson's suggestion, that in principle it might be possible for each one of the five modes he describes to become dominant at each stage. From our point of view, however, we have found it more correct to assume that the possibilities existing for the "solution" of a phase conflict are somewhat different from one phase to another, that at each phase certain motility patterns are more focal than others and that the "solution" arrived at will be predetermined by the nature of the motility pattern in question. In this respect we are in a way in agreement with the classical psychoanalytic approach, which maintains that the child has different defense mechanisms at its disposal during different phases of psychosexual development.[1]

Our main departure from both Erikson's and the orthodox psychoanalytic model is our consistent emphasis on respiration and postural patterns, on the necessity of an integration to take place

---

[1] Beres (1956), for instance, writes: "The ego, in its different phases of development, approaches the conflicts related to reality and to the instinctual drives with different defenses and different solutions, according to the specific phase..." (p. 209).

Figure III

Schematic Survey of Pregenital Developmental Stages
Modes of Activity

Stage	Focus of pleasure	Aggressive manifestation	Emerging phase modality
Early oral	sucking	crying (diffuse rage)	receptive-responding
Late oral	chewing	biting	receptive-self-providing
Early anal	defecating	expelling	eliminative-selv-discovering
Late anal	shaping (evacuation control)	defying	eliminative-self-deciding
Early genital (male)	making (copulation-like play)	aggressive-intruding	introducive-self-expanding
Early genital (female)	making (copulation-like play)	aggressive-embracing	embracing-enclosing self-expanding

between respiration and various activity patterns in order for
these patterns to give complete release and satiation, and on the
possibility of postural modifications emerging as a pseudo-
solution to the integrative tasks confronting the child. Although
the primary aim of these solutions are to resolve conflicts, they
will always stabilize and freeze the conflict on a latent level
and probably always have repercussions on the child's respiration,
making it less adaptible to later needs and challenges. In relation
to each developmental phase we have indicated the possibility of
two types of pseudo solutions - the one going in a hypertonic, the
other in a hypotonic direction. Although these solutions in one
sense are opposite ones, we may also conceive of them as vicarious
in the sense that the one may replace the other, and that a rapid
shift between them may be considered an additional particular type
of solution.

Which type of solution that actually will develop in a
certain stage probably will depend not only on the nature and the
intensity of the frustration and deprivation confronting the child,
but also on the solutions arrived at at earlier developmental
stages. As stated, already immediately after birth infants show
differences in their respiratory patterns. And later on, the
defensive solution arrived at will probably in most cases not only
have an internal stabilizing function, but also an external one,
furthering the child's immediate adaptation to environmental demands.

In congruence with Erikson we may suggest that the different phases of development and the modalities being established
in these phases, will be decisive for the formation of certain ego
qualities: the oral phase as to whether basic trust shall characterize the growing ego, the anal and the early genital phases, as to
whether autonomy and initiative shall become intrinsic ego qualities.
Enduring defense solutions will affect the development of these
qualities. In fact, we may in such instances probably talk about a
conflict-charged ego-trust, a conflict-charged ego-autonomy, etc.

A person's character - the sum total of the modes of
reacting which are specific for a given personality - we may assume,
will be affected by his hereditary constitution and by the modalities

developed at the various psychosexual stages. Normally a person's
character will represent a broad adaptation and organization of
early phase modalities, an organized whole developed on the basis
of his total inner resources and external experiences. In the
case of early defense modalities, these will always exert a specific
influence on a person's character formation, making it less plastic,
and to a certain extent, determine the person's later interaction
with his environment, and thus, both directly and indirectly,
contribute to the distinctiveness of the adult personality. Certain
defense modalities, may in fact, work very much as a catalyzing
nucleus of the total structuring of the personality. The designations
oral characters, anal characters, phallic narcissistic characters, and
so on, imply in our opinion, just such a dominant organizing effect
of a particular type of phase modality.

Following psychoanalytic conceptions of pregenital character
traits we want to emphasize that these traits are not psychopathological in the usual sense of the word, but represent a neurotic reaction
basis which may or may not develop into a psychopathological state.
According to psychoanalytic theory most persons' characters stem from
pregenital psychosexual fixations, as well as from sublimation of
pregenital urges. By sublimation is meant the withdrawal of energy
from a pregenital impulse pattern in favor of its cathexis on other
impulses which make possible an adequate discharge of instinctual
energy. While some psychoanalysts, e.g. Reich, seem to consider
"genital primacy" a prerequisite for sublimations, others, e.g.
Kris, referring to observations on children's play, maintains that
sublimations do take place in pregenital phases too.

When stating that character traits may stem from pregenital fixations, it is sometimes meant that the character trait
in question is a continuance of infantile libidinal urges (Abraham),
and sometimes, that it is a reaction formation against still operative
pregenital urges (Fenichel).

Our own point of view is closer to Fenichel's than to
Abraham's on this latter point, while on the former we feel closer
to Kris than to Reich.

In viewing postural dissolutions of infantile responses as reaction formations we make, of course, an energetic assumption, an assumption that the blocking of infantile responses will not extinguish their energetic basis, but rather change these into self-perpetuating drive systems seeking expression in indirect and disguised ways, in ways congruent with the postural solution arrived at.

It follows from this viewpoint that early defense patterns, themselves functioning as reaction formations, may give rise to new conflicts, to new ambivalences, and to secondary postural changes and secondary reaction formations. In short, we may be confronted with possible hierarchical arrangements of postural patterns stemming from the same tensional source. But, as mentioned, we may also take account of the possibility that fixations may be spontaneously dissolved, that sublimations may take place through the child's interaction with its environment, and also that the postural defense of several conflict nuclei existing in the same individual may overlap to a certain extent and eventually merge into more comprehensive patterns. For instance, we have touched upon the fact that a postural defense against both anal-expulsive and genital-intrusive responses may involve the pelvis and the leg muscles. Although somewhat different muscle groups may be primarily involved, the structuring of leg tensions in the case of genital conflicts, most probably will be influenced by whether unresolved anal leg-tensions exist or not. In principle, we may assume the same type of structuring to apply to most body parts and that manifest postures usually are multiple determined. But, as stated, not only would we expect antecedent conflicts to influence the postural solution of later ones, but also the child's subsequent integration with its environment to influence the postural solutions earlier arrived at. This latter point brings us over to the hypothesis launched by Deutsch and others, that not only environmental factors may play an important role in this connection, but also intraindividual cohesive synthe-sizing forces of a maturational nature.

Deutsch (1947) states:

"Certain postural attitudes that have developed independently
become finally integrated with each other, and when the
personality is fully developed they become consistent with
one another, presenting the characteristic expressive acts
of the adults. Behind the specific expressive nature of a
posture or movement is invisibly enacted the whole immanent
latent organized structure of the integrated system."
(p. 198).

And in another article (1949), he states:

"All postures of the different parts of the body are attuned
to each other, and the change of one partial posture leads
to rearrangements of the total configurations... The configurations of the posture are bound to be characteristic
for the person in question and depend upon the firmness of
the synthesization of the postural pattern within the
personality." (p. 62).

One important aspect of this synthesizing process is
probably represented in the different types of bodily role-play
going on through childhood, through role identification with
animals, through symbolization of bodily parts, e.g. the identification of various fingers with family members, and the left and
right part of the body with "active" and "passive", "strong" and
"weak", etc. Probably these processes represent a salient feature
of the developmental phase interlinking the early and late genital
stage of development.

To summarize, our view is that a child in the course of
its psychosexual development will go through various phases characterized by the emergence of specific modalities. If the child is
given sufficient emotional support the various modalities will be
"grown out of" and "grown away from"; the emerging modalities will
be assimilated in the ego structure giving rise to the emerging of
new modalities and to the ego's progressively more complex level
of psychic functioning. Given an overwhelming conflict, a defense
pattern will come into play and function as a reaction formation

against these impulses which the child has come to consider dangerous.
In this way a developmental fixation will be created, which may become
permanent, or temporal, depending upon the subsequent interaction
pattern between the child and its surroundings, and, not least, upon
the strength of the child's congenital vitality and synthesizing
capacities. The dissolution and transformation of fixations, we
think, may take place in childhood as well as in adults, "spontaneously" as well as in therapeutic sessions.

Looked at from the point of view of therapy, personality
changes may be considered to take place within the framework of
learning, although not in toto. Concentrating on repetitious fragmentary responses, on responses unintegrated with breathing, the
problem confronting the therapist is not so much the patient's lack
of learning, as postural features making learning inpossible. A
therapeutic approach aiming at the reestablishment of the postural
basis for learning would be insufficient if not followed up by
learning, as would a respiratory approach aiming at learning, without first establishing the postural prerequisites for learning. In
this latter area a psychological approach might frequently be much
more effective than a physiotherapeutic one.

EXPERIMENTAL STUDIES OF POSTURAL DYNAMICS

In the last chapter we launched a general developmental theory concerning postural configurations. A basic assumption in our theorizing was that postural and muscular tension expresses psychic tension and conflict, and that a person's developmental history in terms of conflicts, fixations and repressions, will be revealed in his present postural picture.

Muscular tensions can be observed and recorded through overt behavior. In the first chapter we dealt with a number of hypotheses concerning the relationship between overt postural patterns and psychological traits. However, postural patterns and motor activities are for the most part invisible. Consequently their measurement requires spesific recording techniques. Most common, is the measurement of the electrical activity going on in the muscles. In this chapter we shall dig into laboratory studies based upon such measurements, i.e. EMG recordings. We will touch upon some of the hypotheses mentioned earlier, but still more basic will be our discussion of the psychological meaning of muscle tension potentials generally.

Traditionally muscle tension has been dealt with in experimental psychology from the point of view of thinking. Several hypotheses have been formulated concerning the muscular-motoric aspect of cognitive processes. According to Jacobson, for instance, much ideational activity is really ideomotor activity. The behavioristic tradition in psychology has stimalted a good deal of theorizing and research in the area of muscle potentials. However, only a small part of this endeavor has been focuced upon personality dynamics proper.

Concentrating on personality dynamics we would like to mention a relatively recent school of thought working with muscle tension as an expression of an organism's general level of behavioral arousal and activitation, i.e. energy mobilization, in an attempt to reduce the descriptive categories of psychology to two basic dimensions, the intensity and the direction of behavior. EMG-potentials have been interpreted as an indicator of an individual's activation level or behavioral intensity. In addition to EMG, skin conductance, EEG, pulse rate and respiration rate has been used in the measurement of the intensity dimension.

Duffy (1957), one of the strong advocators of the concept of activation, points out that the degree of general internal arousal usually correlates fairly closely with the intensity of overt response, but that a discrepancy between the two may be introduced by the intervention of inhibitory processes. "A phenomenon", she states, "which has not received the degree of attention to which it is entitled."

As regards the generality of muscular tension, Duffy has the following to say:

"It appears that, on the whole, skeletal-muscle tension in one part of the body tends to be positively related to that in other parts of the body, though the relationship between the tension in any two areas may not be very close. Parts of the body more remote from each other, or more widely differentiated in function, yield tension measures which are less closely related than those which are closer together or functionally more similar. When tension measures taken from different parts of the body, recorded during different tasks, or made at widely separated intervals in time, nevertheless show a significant positive correlation with each other, it must, however, be concluded that there is at least some degree of 'generality' in skeletal-muscle tension." (p. 271)

In postulating 'generality' in muscular tension Duffy cites some earlier studies of her own of nursery school children showing marked individual differences in grip pressure, and a significant correlation between the grip pressure on one occasion and that on another, and during one task and that of another. Of course these studies tell us more about consistency of tension over time than about consistency of tension of various muscles at any one time. The same is true of a couple of other studies referred to. Most revealing is a study of Lundervold (1952), who found that subjects characterized by high level tension in (some) skeletal muscles were more responsive to a wide variety of stimuli than subjects with lower tension. Thus 'tense' subjects, as compared to 'relaxed' ones, were found to show more activity in the muscles when external conditions were changed, as by an increase in noise, the lowering of the room temperature, or the introduction of stimuli intended to cause irritation. In the 'tense' subjects there was not only more activity in the muscles participating directly in the movement task at hand (typewriting), but also in other muscles not directly involved. This last finding points to a generality

factor, although the number of electrodes employed was too small and
their placement too restricted (the arms and the upper part of the
body) to warrant any definite conclusion about a 'generality' factor.

The observation that the pre-stress level of muscular tension
related positively to tension increase during stress is a remarkable
observation - and we will return to this question in a later section.

Granted for a while a 'generality' in skeletal muscle tension,
what sort of factors are responsible for the degree of tension present
and what does a certain tension level stand for in terms of psychological
functioning?

As regards the latter question, Duffy (1957) as well as Malmo
(1957) equates tension level with activation level and behavioral intensity; i.e. the intensity of drives, motives, and emotion characterizing an individual as a whole at a given time. Fisher and Cleveland
(1957), touching upon the same topic, the physiological reactivity of
the body exterior (here too with an implicit hypothesis of generality),
suggest that it is "correlated with alertness, striving and active
expressiveness."

Several factors are responsible for variations in the degree
of arousal, Duffy maintains. Such factors include hormones, drugs,
physical exertion, genetic conditions, and the present external situation, i.e. its incentive or threat value. Consequently, if all factors
affecting the level of arousal except situational ones are held constant,
measurement of the degree of arousal may afford a measure of the
'motivating' or 'stress' value of a given situation.

Specific attention is payed by both Duffy and Malmo, to the
effect of various degrees of activation upon performance, under and
over-arousal being considered inferior to an intermediate, optimal level
of activation. That the average EMG level (from pronator teres and
pronator quadratus of a non-active left arm) do correspond closely to
the degree of incentive in an auditory tracking experiment is shown in
a study by Stennett (cited by Malmo, 1957), while Wallerstein (also
cited by Malmo, 1957), found frontalis muscle tension, in subjects
reclining in a comfortable bed, listening to verbal material presented
by a tape recorder, to be positively related to the subject's reported
interest in listening.

The optimal level of activation is probably related to the task to be performed, whether it is new or habitual, perceptual or motoric, and to characteristics of the individual, i.e. to his excitement tolerance, his ability to function under high levels of tension.

A very fascinating question is raised by Malmo (1959) in a recent article on activation: Does an organism's activation level exist independently of the present environmental stimulating conditions? Is it possible to talk about the deprivations of certain biological needs (e.g. thirst) resulting in a monotonically increasing activation level but the increment being latent unless and until the organism is observed under appropriate environmental conditions?

The relationship between manifest and latent muscular tension and the interplay between characterological and situational factors introduces some very important viewpoints directly related to our own developmental model. We will return to these questions, too, in later sections.

EMG-potentials as an indicator of behavioral conflicts.

Differences in action potentials might refer to other processes than to behavioral arousal generally. This is indicated in a study by Shagass and Malmo (1954). Comparing day-to-day EMG recordings from a patient in psychiatric treatment (mean tension per interview integrated from right neck, right forearm extensors, right peroneal muscles, and forehead) with routine notes on behavior and mood by ward nurses, a high positive correlation was found between a high (respective low) over-all tension and a depressed (respective cheerful) mood.

In a study by Whatmore and Ellis (1959), comparing EMG measures from depressive patients and controls it is likewise concluded that depressive patients (in a relaxed supine position) exhibit significantly higher (invisible) general muscular tension.

To conclude from these studies that a depressive state is accompanied by a higher energy arousal than a cheerful mood seems un- warranted. More likely is an interpretation of high EMG potentials as indicating a conflictual state of mind, a large amount of energy under depression being directed inward in order to bind outgoing impulses.

Such an interpretation fits well the conclusion drawn by Gottschalk et al. (1950) on the basis of a comparison between simultaneous electromyographic and psychoanalytic observations. They state:

"In general, comparing the psychological data with the electromyograpic data showed that when the patient's defenses against hostile impulses allowed adequate and otherwise socially permissible expression of these impulses, the patient's muscle tension was decreased, but when psychological tension was aroused, through intra-or-extra-psychoanalytic experiences, because of inadequate means of coping with hostile impulses, the patient's muscle tension was increased, i.e. inhibition of effective aggression led to increasing muscle tension, expressed verbally by the patient as the complaint of being 'muscle-bound' or 'strait-jacketed' at such times." (p.741).

A study by Malmo et al. (1952) gives further support to the hypothesis of a positive relationship between EMG potentials and behavioral conflicts. In the study referred to, higher muscle tension was found with motor conflict in the absence of overt action, than with conflict-free overt action.

This last point brings us over to a problem confronting EMG studies, the problem of differentiation between overt movement and the static tensional state of a certain muscle group from which data are being gathered. Although muscle potentials usually are higher during muscular movement than during non-movement, this is, as just mentioned, not always the case. However, the question of movement _versus_ tension is an ever recurrent one.

An experimental study by Malmo et al. (1950) illustrates the problem. In this study it was found that mental patients with head complaints reacted to a non-specific stress situation by higher neck muscle potentials than patients without such complaints. One question immediately confronting the experimentor was how to interpret this finding. Did the higher neck tension potentials in the first group indicate a higher tensional state or only a greater frequency of head movements? In order to control for this latter factor, head movement scores were computed for all subjects. Comparing these scores for the two groups no positive relationship was found, and it is concluded that "the muscular contraction associated with head complaints are distinct from contractions leading to those overt head movements which may be observed behaviorally." (p.236).

The experiment just referred to brings us back to our initial
question: The generality of muscular tension. Does the higher general
tension found to characterize depressive states hold up for each and
every body region? Does the higher neck tension found in the head-
complaining patients indicate a higher generalized tension in this
group of patients? Neither Whatmore nor Malmo do think so. What was
found, Malmo suggests, was a specific, high neck-localized tension increase in the head-complaining patients. In Whatmores study, four body
regions were selected for EMG recordings, namely the forehead, the jaw-
tongue area, the forearm and the leg. Comparing a group of extremely
depressed patients with a group of normals matched for sex and age, they
found higher tension in all four regions, but when replicating the study
on a group of depressed patients not exhibiting the extreme motor retardation of the first group, they found a higher grand mean than in the
control group - although significant differences were found in the jaw-
tongue area $(p < .01)$ and in the forearm $(p < .05)$ only.

The results of a couple of later studies by Malmo et al. (1955)
in fact all point in the direction of a fairly high degree of tentional
specificity. In one of these studies, a factor analysis was computed
of tension measures from six muscle groups during non-specific pain
stimulation. No factor of generalized muscle tension appeared. Neck
and bilateral-forearm tension (right and left forearm extensors and
flexors) fell into one single factor, while forehead tension and
measures of movement irregularity fell into another. In another study
by the same researchers, of the effect of listening and talking in
relation to EMG (from frontalis, chin, neck, and left and right forearm
extensors) only a single factor could be extracted. This involved
frontalis and chin tension and motor irregularities, while neck and arm
tension fell outside and seemed independent of each other. In discussing the results Malmo et al. emphasize that forehead tension and
movement irregularities seem to be the best discriminator between psychoneurotic patients and normals, and that frontalis-muscle tension may
provide an index of something general in the complex phenomenon referred
to as psychoneurosis.

Does localized tension indicate specific conflicts?

Granted that a positive association exists between psychological conflicts and muscular tension, and that various individuals exhibit different tension patterns, the next question confronting us is whether an association exists between specific tension patterns and specific psychodynamic conflicts.

Some very interesting findings concerning the relationship between psychodynamic and tensional specificity are presented in a study by Shagass and Malmo (1954). The study, focused on three patients only, tries to answer the question whether the discussion in psychotherapy hours of certain dynamic topics or themes is correlated with tension increase in specific muscle groups in the patients.

The first case presented involves a minute-to-minute comparison over 9 interview sessions of the relationship between 'hostility' and 'sex' themes in the interview process and EMG measures from the patients' right neck, right forearm extensors, right leg, and forehead. The results reveal that 'hostility' content is associated (and only associated) with increased right forearm tension, 'sex' content only with increased right leg tension. Furthermore, comparing EMG potentials from the right forearm and overtly hostile and covertly hostile state- ments about the same objects (mother, sister and sibling surrogate figures) it was found that the latter type of statements tended to be associated with more arm tension than the former type.[1] Finally, com- paring EMG potentials and statements referring to movements and to feelings or opinions only, it was found that leg-tension was unrelated to statements involving movements, and arm-tension, inversely related to movement. The relationship discovered between increased forearm tension and hostility was found positively not to be a function of hostility statements involving more movement references than others.[2]

----

1) We are here reminded of Gottschalk's statement "that when the patient's defenses against hostile impulses allowed ... expression ... the patient's muscle tension was decreased ...".

2) It is likely that overtly hostile statements will be positively associated with statements involving references to movements. Thus, we would expect that as a patient moves in psychotherapy from covert, abstract and indirect expressions of hostility (e.g. toward parental figures) to more overt expressions, a revival of concrete events in- volving movement references will occur. However, since more overt expression of hostility probably will be associated with less tension, it is not surprising that arm tension was found inversely related to movement references.

Although less thoroughly analysed, the case material presented
for the two other patients substantiated the main findings from the
first case.[1]

How come that specific ideational content get associated with
increased local muscle potentials? That thought of movement leads to
tension in the muscle which is imagined to move has been demonstrated
experimentally by Jacobson (1938) long ago. However, Jacobson worked
with conscious imagining activity only. The fact that muscle tension
increased just as much (sometimes even more) when the sucjects statements referred to non-movement as to movement, discourage an explanation in terms of <u>conscious ideomotor action</u>.

Shagass and Malmo end up by suggesting that "muscle potentials
are an indicator of central nervous system activity underlying conflict
in the sense that increased muscle tension reflects ineffective
resolution of conflict within the central nervous system." Arguing in
favor of such a view, they state:

"Conflict involves simultaneous presence of opposing action
tendencies (impulses). These action tendencies must be mediated
through central neural processes. When opposing tendencies are
present together, some form of central neural interaction must
occur, which provides the mechanism for resolution of conflict.
The simplest possible central interaction would be mutual inhibition of opposed processes. Such mutual inhibition should
eliminate peripheral manifestations of conflict (muscle potentials
in this instance)." (p.312).

Concentrating on their empirical observations, they continue:

"Emotional conflicts of interest to the psychiatrist are usually
of long duration. Most of the time, the opposed impulses are
presumably in some sort of balance. In the interview the
balance may be upset by discussion of material which temporarily
increases the strength of the unacceptable impulse. The opposed
impulse must be reinforced quickly in order to restore balance
and prevent overt action and perhaps awareness. This happens
automatically through processes which may be allied to repression.
However, unless the opposed neural processes completely inhibit
one another within the central nervous system, increased tension
would be expected in the appropriate muscles. This hypothesis
accounts for the symbolic meaningfulness of present correlations.
It accounts for the fact that tension and theme were statistically
but by no means invariably, associated, since the effectiveness
of central inhibition must vary (as does the effectiveness of
repression). It accounts also for the higher muscle tension
associated with strongly conflictual, as opposed to weakly con-

1) Consequently also some central features of the hypotheses
   presented in chapter I.

flictual or neutral themes, since any action tendencies
associated with the latter should be easier to inhibit centrally."
(p. 312).

        Shagass and Malmo's hypothesis does not give any account of
why specific themes seem to be associated with specific muscle groups,
except suggesting that in the case of inefficient central conflict
resolution "increased tension would be expected in the appropriate
muscles". It is somewhat unclear what is meant by this statement.
Does it mean that the motor muscular innervation of an "unacceptable"
impulse will be inhibited, controlled, and balanced peripherally by
the simultaneous innervation of antagonistic muscles? Or, does it only mean that increased peripheral tension will emerge without this
tension having any functional purpose?
        In what follows we will discuss these viewpoints a bit further.

EMG-potentials as an indicator of nervous overflow.

        The conception of EMG potentials as an indication of nervous
overflow is salient in a couple of articles by Malmo et al. Thus, in
one article (1956) we are told:

> "How does motor conflict produce this increase in tension?
> Neurophysiologically, increased tension must mean more neural
> impulses coming down the final common path. Very briefly, the
> hypothesis is this: The increase in the number of neural im-
> pulses in motor pathways is due to central neural interaction.
> Motor mechanisms for approach compete with similarly organized
> mechanisms for avoidance. This competition produces a two way
> interference which evolves impulses which are not part of any
> organized motor pattern. These extra or disorganized impulses
> going to the arm muscle would raise its tension. The same
> hypothesis may be applied to conflict over hostility where the
> opposing instigators may be--say--aggression versus affiliation
> or deference. A similar principle is applicable also to sexual
> conflicts." (p. 271).

        In another article, Malmo et al. (1954) elaborate on the same
viewpoint as follows:

"How do we interpret the muscle-potential increments associated
with motor conflict? In general terms, we may state that these
potentials reflect incomplete central inhibition. Two or more
incompatible neuronal systems are simultaneously activated.
Suppose we consider the simplest case: two such systems A and B.
They are incompatible because there is mutual interference with
the motor facilitation required for smooth coordinated action.
Either A or B alone can gain the necessary facilitation, but
together there is interference. This is the neural paradigm of
conflict. Now if B is quickly canceled, A immediately gains
control and organized motor action proceeds without interference.
In this case conflict is brief...Let A represent flexion (withdrawal) and B extension...(approach)...Complete inhibition of both
A and B...the subject is relaxed. Complete relaxation of this
kind would be produced by simultaneous stimulation of inhibitory
nerves to both flexors and extensors. This may be considered as
the base line...Deviations from such a base line are of three
kinds. (1) Excitation of A combined with inhibition of B will
lead to uninhibited withdrawal... (2) Inhibition of A combined
with excitation of B will lead to extension (approach)...(3)
Partial inhibition of A and B will result neither in relaxed
acceptance of stimulation nor in a co-ordinated movement. Instead,
we should expect to find 'purposeless' contraction of flexors and
extensors, that is, motor conflict." (p.182).

In the first quotation Malmo et al. refer to the recorded
increment in EMG potentials as an expression of "disorganized impulses",
in the second quotation, as "purposeless" neural activity. Behind
their reasoning is the conception of EMG potentials as nervous overflow.

EMG-potentials as an indicator of peripheral inhibition.

In a later article Malmo (1957) refers to Cameron's observation
of body movements in anxious patients, such patients usually showing
restlessness and constant movements, movements not of the open, wide,
flung-out type found in manic patients, but movements remaining within
the body silhouette. Here he states:

"On the surface, this appears incongruous with the notion of
weakened inhibition. However, we may account for this
constrained appearance of inhibition by suggesting the sub-
stitution of less efficient mechanisms of inhibition for the
one which has suffered impairment. It may be, for example,
that anxiety patients compensate for weakened autonomous
mechanisms by calling on voluntary motor mechanisms (i.e.
the pyramidal motor system). For example, in the absence of
sufficient control from autonomous inhibitory mechanisms, the

anxiety patient may avoid loss of motor controll through
co-contraction of antagonistic muscles." (p.284).

It follows from this viewpoint that peripheral expression
of conflict far from being "purposeless" or an "overflow phenomenon",
has a specific conflict-resolving purpose. That is, Malmo suggests
that in the case of an impaired central autonomous inhibition, a
peripheral mechanism, operating within the pyramidal motor system,
may come into play.

At this point we would like to emphasize a difference between Malmo's and Braatöy's (1947) conception of motor defences.
In agreement with Malmo, Braatöy too conceives of peripheral tensions
as purposeful phenomena, but in contrast to Malmo, the defensive
function, not of the pyramidal but of the extrapyramidal motor system
is especially underlined.

Somewhat simplifying we may formulate Braatöy's viewpoint as
follows: Muscular tension may be either of a tonic or of a phasic
nature. While the phasic system is mostly influenced by nerve ter-
minals from the pyramidal system, have shorter latency, shorter duration
and higher specificity, so is the tonic system mostly influenced by
nerves from the extrapyramidal system, with longer latency, duration,
and diffuseness. And while phasic responses are mostly emitted from
the motor region of the cortex, so are tonic responses mainly controlled by the reticular formation in the lower brain stem and mainly
under the influence of interoceptive and proprioceptive stimulation.
For instance, it has been found that the decorticate mammal (rabbit,
cat, dog) shows approximately normal postural activities. Likewise,
that changes in the head position of a cat (changing the neck posture)
produce great changes in the postural segment of its foreleg muscles,
while the movement segment of the very same muscles (m. triceps) is
practically non-affected. In other words, proprioceptive impulses
seem to produce postural changes, postural changes to produce proprioceptive impulses, and so forth, until a certain balance is attained.
One is here reminded of Sherrington's view that no other afferent fibers
than those of the tonic muscle itself are essential for the exhibition
of tonus. And one is led to think that postures once initiated under
certain circumstances may have a sort of reflexory, self-perpetuating

function, and that the somatosensory impulses reaching the brain stem to a large extent are dependent upon efferent impulses stemming from this very layer.

We find somewhat similar thoughts behind Allports (1955) conception of a set as a readiness to respond in certain directions and as represented by increased excitation in given muscle groups. Describing the development of a set, Allport states:

"Some initial stimulus in the recent or more remote part has led via the central nervous connections to a motor response, as for example, in limbs, hands, vocal organs, or the muscles of sensory accomodation. Proprioceptive stimulation in the muscles, tendous, or joints arising from that response starts impulses in afferent neurons leading back to the central nervous system. From there impulses produced or reinforced by the excitation flow out again into the musculature of the same effector organs. The contractions, so produced or reinforced, again set up proprioceptive backlash to the central nervous system - and so on. There is thus a circular process that can maintain itself for indefinite periods of time, that can, in a sense, become independent of time, until the internal condition of the organism or the environmental situation, changes." (pp. 227-228).

How do the above considerations fit in with the view that tonic or postural patterns might serve an impulse-defensive function, i.e. that they might inhibit phasic responses?

Two principles of explanation are available, the one focusing on central, the other on peripheral processes. The latter one takes its point of departure from the assumption that every phasic response presupposes a certain postural substratum, that if this substratum is not present, the response will be inhibited in spite of facilitating external conditions. The former one emphasizes the close neural connection which seems to exist between the reticular formation and the limbic system (i.e., the layer forming the border or "limbus" around the brain stem), and recent neurological findings suggesting that within the limbic system there are regions or centers (e.g. one of them the hypothalamus) which when stimulated evoke reaction patterns like anger, fright, joy, laughter, and sorrow. Since the limbic centers seem to have little link to the neopallium and the mesopallium (e.g., it has been shown in animals that bilateral lesions in various segments of the limbic system result in affective changes but not in measurable de-

terioration of intelligence), but close linkage to the archipallium,
it is suggested that the reticular formation may exert a releasing as
well as inhibitory influence on the affective centers within the limbic
system. That is, in a way, that parallel to a postural substratum
being necessary in order for a certain response to be elicited, a
certain activation is necessary for a limbic center to be aroused, and
that parallel to a postural inhibition of a certain response we may
talk about a reticular inhibition of a certain limbic center.

Granted the above train of reasoning, and one additional
assumption, namely that the inhibitory and facilitatory functions of the
reticular formation vis-a-vis the limbic system are interdependent upon
the pattern of proprioceptive (somato-sensory) impulses reaching the
reticular formation, we end up with a viewpoint stressing the intimate
interaction between central and peripheral inhibitory processes. In
opposition to Malmo's assumption, 1) that opposing neural processes
and their inhibition primarily take place centrally (by the hyper-
polarization of cortical cells), and 2) that only if a resolution on
this level does not work will "motor pathways" and muscle tension be
involved, we end up with the proposition that no sharp distinction
exists between central and peripheral resolutions, only between various
types and patterns of "central-peripheral" solutions -- involving various
muscle groups and various limbic centers, respectively. This, we feel,
is the essence of Braatöy's theoretical position, and it is also
indirectly the position for which we made ourselves spokesmen in the
preceding chapter.

In emphasizing that feedback from proprioceptive end organs
within muscles represents an important mechanism in impulse inhibition
and repression, physiotherapeutic observation of impulse mobilization
and the breaking through of impulses through peripheral (postural-
muscular) manipulations may be cited (Braatöy 1947; 1954), as well as
more general neurophysiological evidence that feedback excitation from
muscles is both a distributor and contributor of excitation and inhibition within the central nervous system. For instance, Gellhorn (1949)
has shown that proprioceptive feedback from one muscle can increase the
reactivity of other muscles, and (1948) that action potentials from a
muscle will be disproportionately increased during stimulation from the
motor cortex if the observed muscle is fixed in a lenghtened position.

In a recent article, Malmo (1959) too suggests that peripheral
tension manifestations may give rise to proprioceptive feedbacks,
although he considers these feedback processes within a rather restricted theoretical framework. Malmo states:

> "From a clinical point of view it seems an interesting specula-
> tion that the patient's localized muscle tension may itself
> actually increase the general activation level... that the area
> of localized muscle tension in the patient acts like tension
> that is induced (i.e. in producing a quite general muscle
> tension)...that the local increase in muscle tension somehow
> produces an increase in the general level of activation, with
> rise in the heart rate and blood pressure, with fall in level
> of EEG alpha, and so on. ...If such is the case symptomatic
> treatment might have significant general as well as specific
> effects." (p.384)

Malmo's assumptions account well for Gellhorn's findings and
the principle of motoric conflicts giving rise secondarily to anxiety
and autonomic changes. However, they leave out of the picture the
question of postural defenses and the very important observation that
the resolution of a localized muscle tension very often is followed
by the displacement of tension to another muscle group.

Braatöy (1952) comments upon this last phenomenon may serve as
an illustration of his theoretical position generally. He states:

> "When we have freed the pains and restriction of movement at
> one spot, the same symptom may reappear at another place.
> Concretely, if the physiotherapist has succeeded in releasing
> the fixed m. latissmus dorsi and likewise in removing by
> massage the painful nodes of the muscles, it not infrequently
> happens that hypertension occurs at other groups of muscles,
> resulting in restriction of movement and attendant myalgia.
> This...can be understood when we realize that a comprehensive
> primitive reaction may be suppressed in different ways - an
> emotional reaction chiefly by gritting the teeth, or by clenching
> the fists, or by holding the breath...Treatment directed toward
> the local symptoms will hardly be effective unless it, at the
> same time, takes into account the more extensive primitive re-
> action. Sometimes the treatment of the local symptoms may be
> successful but the underlying movement-complex becomes firmly
> fixed at another place. Then symptoms arise at that place in
> due time. On this basis one can understand why the results of
> physiotherapeutic treatment of nervous affection may be so
> problematic unless the treatment can remedy two things at the
> same time. The local symptoms and the more comprehensive
> primitive reaction (the emotional factor). In ordinary medical

and physiotherapeutic tradition, the first phenomena are
attacked, while the others are disregarded. In traditional
psychiological theory and in psychotherapy, the situation
is often the reverse." (p.221).

Let us briefly sum up the three viewpoints on localized
tensional increase mentioned so far.

The first one states that tensional increase in a localized
muscle group is an overflow phenomenon, that it is purposeless and an
expression of disorganized impulses. Thus the localization of the
muscle group involved may be considered somewhat arbitrary.

The second one also states that it is a sort of overflow
phenomenon, but that it is purposeful in the sense that it represents
attempts at curbing organized motor responses peripherally by the co-
contraction of antagonistic muscles. The localization of the muscle
group involved is determined by the motor responses being defended
against.

The third viewpoint states that localized tension increase
represents a reinforcement of a postural (tonic) defense threatened
to be overrun by the reinforcement or mobilization of defended against
emotional reaction patterns (comprehensive primitive reactions). Here
too the localization of the muscle group involved would be determined
by the responses being attempted warded off.

Comparing the latter two explanations, a couple of differences
appear. In both cases we should expect tension increase in relevant
muscle groups, the more a threatening response threatens to emerge.
However, in the former case we are lead to assume that the tension increase in question is caused by phasic impulses only, while in the
latter case, the increase would be assumed to be caused at least partly
by tonic or postural mechanisms. Furthermore, in the latter case we
would assume the localized tension increase to correspond to a habitual
"dystonic" tension in the muscle group in question, the situationally
induced tension-increase being partly a function of the muscle's
habitual tension level. This last assumption follows from the view
that "repression" of conflicts (giving rise to characterological
defensive solutions) are always a matter of central and peripheral
processes, and that an individual's sensitive spots or conflict-areas
will be expressed in postural patterns preventing the emergence of the

appropriate postural "substrata", facilitating the "triggering off"
of the threatening impulses. In short, we should expect the more
situational tension-increase in a certain muscle group the higher the
base-line tension of the muscle group in question.

Focusing on this latter proposition one may certainly say that
the same prediction could be made from the assumption of peripheral
tension being an overflow phenomenon: The more the habitual overflow,
the greater the increase in overflow when threatening impulses are
situationally activated. However, a difference would again be present
in terms of the phasic (movement) versus tonic (postural) nature of
the habitual tension pattern.

In order to test empirically which of the two explanatory
principles are the most valid, a method by which it was possible to
distinguish between phasic and tonic EMG-potentials would have been
most helpful. However, such a method is unknown to us. Another
complicating facts is that tonic innervations stemming from the extrapyramidal system have much greater diffuseness and longer duration,
so that their action potentials are difficult to ascertain electro-
myographically. It has been claimed, for instance, that postural
muscles may get quite tense, i.e. be found contracted by palpation,
without showing corresponding changes in action potentials.

Are situational changes superimposed upon characterological postural patterns?

Shagass and Malmo's hypothesis regarding the correspondence
between dynamic themes and localized muscular tension are directed
toward situational changes only. The relationships found are treated
mainly as interesting empirical observations.

Starting out from the assumption that increased tension in
certain muscle groups during psychotherapy represents attempts
(unconscious) on the part of the patient to inhibit or repress
'threatening' impulses, and that such an inhibition might be either
a matter of impaired (or insufficient) central inhibition (Malmo's viewpoint) or a matter of a reinforced central-peripheral inhibition (central
and peripheral processes always interacting whether impaired or not),
the next question confronting us is the relationship between 'situational'
and 'characterological' tension patterns.

If for a while we assume that situational changes in EMG-potentials get superimposed upon characterological patterns, we may argue as follows: The more conflictual an individual's relation to certain impulses, i.e. the higher his base-line tension in certain muscle groups, the more easily will he become upset by situational factors touching upon the underlying conflict, i.e., the greater the increase in action potentials from the related muscle groups. Turned around, we may suggest that the higher the habitual tension of a certain muscle group, the greater the tensional increase when conflicts, peripherally "resolved" in the habitual tension, are mobilized in one way or another (through pictures, therapeutic discussions, or everyday life events).

The above proposition gets some indirect support from Lundervold's (1952) findings-previously referred to--that 'tense' subjects, as compared to 'relaxed' ones, show more muscle potential increments when the external situation is changed, by the increase in noise, by the lowering of the room temperature, and by the introduction of stimuli intended to cause irritation.

Further support seems to be present implicitly in Shagass and Malmo's (1954) empirical data, although not attended to by the investigators themselves. As previously mentioned, the investigation in question was focused on therapy recordings from three cases.

Their second case study is based on a fairly directive, probing interview of a 34-year-old woman. If the proposition stated above is correct we should here expect to find the greatest tension increase in a "theme-relevant" muscle group, the greater the base-line tension of the muscle group in question.

Table 1 presents some of their empirical data.

In the case under discussion, Shagass and Malmo found the left forearm extensor only to be significantly associated with "hostility" content, and the right leg only with "sex" content. What is remarkable is that both these muscles are comparatively tense irrespective of content. Both right forearm extensor and flexor show tension-increase during "hostility" content, but being farily low in base-line their tension increment is also fairly low.

Table 1

| | Median Amplitude in Microvolts | | | |
| | Hostility Concent | | Sex Content | |
Muscle area	Present	Absent	Present	Absent
Right leg	38.0	39.5	_57.5_	_36.9_
Left forearm extensor	_82.5_	_65.8_	72.5	74.2
Right forearm extensor	8.4	8.0	6.7	8.8
Right forearm flexor	9.6	8.0	9.3	8.5
Forehead	13.5	13.8	13.0	14.6
Neck	16.9	13.7	13.6	14.7

Turning to their third case study (their first study doesn't present relevant data) we find the same trend. This study is focused on a directive interview with a 17-year-old Jewish man. The patient had recently become aware of death wishes toward his father and "father content" was here used as hostility criterion. Since "father content" was found significantly associated with the right forearm only, we should consequently expect the right arm to show higher base-line tension than the left.

Also included in the table (table 2) are EMG responses to "College failure" - the subject had recently learned that he had failed his university exam, that his possibility for obtaining a scholarship was gone--which necessitated his immedate choice of an occupation--a complex of very frustrating experiences.

Table 2

| | Median Amplitude in Microvolts | | |
Muscle area	Father	College failure	Nonsensitive
Right forearm extensor	_11.3_	_17.5_	_4.3_
Left forearm extensor	4.6	7.5	1.5
Left leg	1.8	4.0	1.5
Right Sternomastoid	22.5	16.3	26.3

We note again that both forearms show tension increase with discussion of 'hostility' themes, the right arm showing the greatest increase corresponding to its higher base-line tension.

The consistently low leg tension corresponds well to the information given in this case that the subject did not exhibit any sexual problems and professed little guilt about masturbation (neither was the interview focused on this aspect of the patient's personality).

Remarkable is the relatively high tension of the right sterno-mastoideus muscle. Tension increase in this muscle was found associated with discussion of mother, sister and sex (median amplitude of 32.5 ɥV as compared to 19.9 ɥV for total interview. "The data seem to reveal an association between sternomastoid activity and discussion of heterosexual figures in this patient, but conclusions are limited by the small number of instances", Shagass and Malmo state.[1]

A fourth case study on the relationship between interview content and muscle tensions is presented in a separate article by Malmo, Smith, and Kohlmeyer (1956). The case studies is a 33-year-old woman.

Table 3

Mean Tension in Microvolts

Muscle Area	Hostility		Sex	
	Present	Absent	Present	Absent
Left arm	21.77	17.02	18.45	18.45
Right arm	35.70	40.80	37.90	43.70
Left leg	10.64	10.95	12.08	9.73
Right leg	3.31	3.54	3.56	3.68

----

[1] Guessing from a developmental viewpoint we would suspect oral conflicts (perhaps centered around the mother) to be at stake. In so guessing we are in agreement with Reich (1942) who suggests that suppression of violent impulses to cry from a very early age corresponds to a tensing of the muscles of the floor of the mouth, the pushing forward of the chin, and the tensing of the lateral neck muscles. That strong impulses to cry might have been particularly mobilized in the patient under discussion seems likely in view of the just preceding events and the fact that the interview was a fairly directive one focused on the patient's actual situation. That the discussion of heterosexual figures mobilized the tension might probably be explained in terms of the mother either playing a dominant role in the initial suppression of crying and or in the relief associated with crying.

Again it is interesting to note (se table 3) that hostility
themes are accompanied by significant tension increase in arm muscles
(in one arm only), and sex themes by significant tension increase in
leg muscles (but again in one leg only).

As we would expect, the leg exhibiting significant tension
increase ($p < .01$) is the one showing the highest base-line tension.
However, turning to the arms our expectations are turned down. Both
arms show a fairly high base-line tension and the right more so than
the left. But instead of finding the greatest tension increase in the
right arm, the opposite turns out to be the case. The left arm only,
shows a significant tension increase ($p < .05$) during hostility content.
As regards the right arm, we find a certain trend toward tension decrease ($p=.18$). Scrutinizing the data further brings out that a
considerably higher variability of tensions seems to be present in the
right as compared to the left arm.

The results of the studies just referred to raise some important
questions: How do we explain the fact that in all the cases discussed
hostility (respective sex) content was associated with tension increase
in one arm (respective one leg) only? Why didn't we find a complete
parallelism between the arms, and between the legs? Furthermore, how do
we explain the fact that the arm showing significant tension increase
was not consistently the same arm from subject to subject? Another
question forced upon us is how to explain the discrepancy noted in the
last experiment, the fact that the muscle group showing the highest
base-line did not show the greatest tension increase.

In the last case, one may certainly wonder if the trend toward
tension decrease in the right arm during hostile content corresponds to the
patient being partially able to express hostility in a tension releasing
manner, or able to do so at least part of the time. Here again we may
refer to Gottschalk's observation that expression of aggressive impulses
corresponds to decreased muscular tension. The apparent sensitivity
of the right arm corresponds to its high base-line tension although it
necessitates a qualification of our earlier assumption that a positive
relationship invariably exists between a muscle group's base-line tension
and its tensional increase during relevant stimulation. Seemingly, the
postural reaction to certain situations may represent a combination of
relief and restraint at the same time.

The question of asymmetry of body reactivity.

We noted above that significant tension increase was not invariably found in the same arm or in the same leg from noe patient to another. In the case of the 34-year-old woman (table 1) it was the left forearm that showed highest tension increase, while in the first case mentioned, it was the right forearm that did so. In discussing their data, Shagass and Malmo (1956) state:

"This difference between the cases may seem of minor significance, but it does caution one not to expect invariable associations between one kind of theme and tension in one location in different individuals. It is more reasonable to assume that each individual will have his own particular pattern." (p. 306).

Seemingly intraindividual differences exist as regards the "reactivity" of the right and the left side of the body. However, this does not, we feel, make it "reasonable to assume that each individual will have his own individual pattern" in the sense that the relationship between conflict-content and body tension location is a completely arbitrary, individualized matter, e.g. that sex conflicts in one case may have a muscular localization identical to hostility conflict in another. Rather we would assume that a specific correspondence exists between localized tensional configurations and conflict content, although within a broader framework each individual will certainly have his own distinctive pattern.

The individual differences in EMG reactivity demonstrated by Shagass and Malmo, bring us over to the question of asymmetry of body reactivity.

Ferenczi (cited by Deutsch, 1947) suggests the left side of the body always to be more accessible to unconscious influences than the right because right-handed people concentrate less interest in it, and in addition, 'right' for many people having the symbolic meaning of 'correct', 'left' the meaning of 'wrong'. Other psychoanalysts too have suggested that each individual, as he learns to differentiate between the right and left side of his body, assigns definite values and meaning to each side. Thus it has been claimed that associations may be formed between the two sides of the body and goodness-badness, strong-weak, masculinity-feminity, acceptable versus unacceptable aspects of the self, etc.

In a series of experimental studies Fisher has pointed out
that intra- as well as inter-individual differences exist as regards
the physiological reactivity in terms of GSR of the right and the
left side of the body. In one of his studies, Fisher (1959) demonstrates that right-handed subjects with good body-image integration
(as measured by the Draw-A-Person-Technique) show relatively greater
left than right reactivity, whereas subjects with bad body-image
integration either show no difference at all or relatively more right
than left reactivity. In another study (by Fisher and Abercrombie,
1958) it is pointed out that right-handed subjects who are secure about
their over-all body images (as measured by reaction to tachistoscopically
presented pictures of mutilated bodies) show relatively greater left
than right reactivity, whereas subjects who are insecure about their
body images either show no difference at all or relatively more right
than left reactivity. In a third study, Fisher (1958) demonstrates
that right-handed subjects who distinguish their left body side as
smaller than the right side (as measured by size judgment through
aniseikonic lenses of the fingers of the right and the left hand) show
greater reactivity of the body side perceived as the smaller, whereas
left-handed subjects and subjects without any clear body image distinction
between right and left body sides, show either no difference at all or
relatively more right than left body reactivity.

The relationship found in the last study between GSR reactivity
and size judgments is commented upon by Fisher as follows:

"An important question has to do with the tendency for GSR
reactivity to be greater on the body side perceived as smaller.
The model for explanation that comes to mind is based on the
idea that the side experienced as larger (and which in right-
handers with a definite GSR gradient is usually the dominant
side) is assigned the function of guiding and controlling motor
responses to problem situations. The executive monitoring function
of one hand in an on-going task is, of course, a well confirmed
phenomenon. The integrity and stability of this set for directive
action may be of considerable psychological importance. Therefore, mechanisms may be developed for preventing surges of excitation which could disrupt or significantly alter the set. This
possibility suggests the hypothesis that disruptive excitation
is "drained off" or channeled into the non-dominant side, where
its effects are relatively less disorganizing. Thus, the non-
dominant side is likely to show relatively greater response to
many kinds of excitation than is the dominant or "large" side.
This formulation implies an optimum homeostatic relationship
between the two body sides." (p. 297).

Fisher mentions a number of studies supporting such a viewpoint. For instance, it has been observed that in situations where right-handed subjects are instructed to perform some simple tasks quickly and simultaneously with both arms, the left arm tends to respond first. Furthermore, it has been observed that when the eyes of a right-handed person are shifting from one fixation point to a new target, the left eye responds first, "snapping" to its new position, as contrasted to the right eye, which proceeds more gradually and in a more coordinated fashion. These studies favor the view that the dominant side is characterized by a stable set which facilitates control but which also inhibits the spontaniety of that side as compared to the non-dominant side. But Fisher's results indicate that asymmetry of reactivity is not a function of lateral dominance only. If that was the case, right-handed (respective left-handed) subjects should invariably show greatest reactivity to the non-dominant side and this was not found to be true.

Of additional importance seems to be the degree to which the individual has developed a right-left division in his body image. Thus, Fisher did not find any significant relationship between lateral dominance and GSR directionality. Although he found a positive relationship between hand dominance and finger-size judgments, it was first when both dominance (laterality) and body image (size judgments) were taken into account that a positive relationship emerged with a certain reactivity gradient in the right-handed group of subjects.

Fisher suggests the following explanation:

"In situations uncomplicated by body-image variables, right-handed dominance may be associated with left directionality of GSR response and left-hand dominance with the opposite GSR directionality. The experiences which determine the clarity of differentiation in the body image of the right and left body sides may occur after the basic laterality gradient has been built up, producing an impact on right or left reactivity within the context of the preexisting laterality gradient. For the average person, the basic process of right-left body image differentiation possibly involves the assigning of different values to the two sides in such a way that attributes of bigness are projected on to the dominant side and attributes of smallness on to the other. If for some reason the division of attributes cannot be made clearly or if bigness is assigned to the nondominant rather than to the dominant side, there may be a disruption or even a reversal of the existing reactivity gradient based on laterality." (p. 296).

From the above hypothesis follows that we should have expected a positive relationship between right side reactivity and the perception of the left side as the bigger one in left-handed subjects. The reason why this didn't show up might be due to the fact that left dominance is almost always far less definite than right dominance and that tension patterns associated with laterality are more irregular in left-handed than in the right-handed subjects, Fisher suggests.

Following Fisher's train of reasoning, the next question confronting us is what sort of values or attributes have to get assigned to the dominant (respective the non-dominant) side of the body in order for this side to maintain its dominance so to speak.

In a second part of the same study, specifically directed toward this problem, Fisher (1959) demonstrates that among right-handed subjects, those who manifest left side dominance in GSR reactivity tend in a Draw-A-Person Test to draw the male figure larger than the female figure significantly more often than do subjects who have no GSR dominance or show right side dominance in GSR. Combining this result with his findings that left side dominance in GSR seems to be associated with good body image integration and security about over-all body image, he ends up with the hypothesis that the existence of a definite GSR reactivity gradient between the two body sides (with the left side showing more reactivity than the right) is based upon a significant body image distinction in the size attributed to the two body sides and that this size distinction is in turn dependent upon whether the individual has learned to discriminate clearly between male and female sex roles in such a fashion that the male is considered to have greater size and strength. To the extent an individual has not learned a stable non-conflictual distinction between the masculine and feminine roles,e.g. if in opposition to the usual cultural stereotype of masculine-feminine size relationships, greater power is attributed to the female than to the male, a great deal of anxious confusion will constantly be present, expressed in body image disturbances and insecurities, and in the lack of a definite and optimum GSR reactivity gradient, Fisher suggests.

Fisher is fully aware that his empirical data is far from sufficient to draw any final conclusion about what sort of attribute-assignments are decisive for differences in body side reactivity. May be we are confronted not with one laterality variable, but with

intraindividual differences in body side reactivity in various parts
of the body. However, Fisher's suggestion that factors related to sex
identification and developmental processes may be involved, attracts
our interest.

Fisher's emphasis on the relationship between developmental factors corresponds to our own earlier conception of postural
synthesizing processes going on continuously through childhood, and
that an individual's final character structure, emerging usually at
the age of puberty, will represent integrations and secondary modifications (through role play identifications and animal identifications)
of genetically more basic impulse-defense configurations. Consequently, in the measurement of physiological reactivity or muscle
tension patterns in adults we might frequently encounter significant
differences between the right and the left side of the body. It is
not at all unlikely that we may be confronted with what Fisher describes as "draining off processes" and "an optimum homeostatic relationship between the two body sides", and furthermore, that such
processes are formed and determined by role assimilations. Our point
is that these assimilations usually represent secondarily postural
modifications, patterned and 'nurished' by the limitations and assets
laid down by the individual's earlier experiences.

Factors responsible for decrement in EMG potentials.

So far we have mainly been concerned with conditions causing
an increment in muscle tension. Obviously, just as important is to
inquire into the opposite relationships.

Not only have the steepness of EMG gradients and factors
responsible for increment in EMG potentials attracted experimental
interest, but also the speed with which an activation returns to its
prior level (the recovery reaction quotient) and the factors responsible for tension decrease. For instance, in a study by Malmo et al.
(1950) on the reaction to strong auditory stimulation, it is pointed
out that it is the after-response following the period of primary
startle-reflex which shows the most reliable difference between anxiety
patients and controls, the former group showing regularly a slower
recovery rate than the latter.

In a study by Malmo et al. (1957) on the physiological effects
of experimentally induced praise and criticism in personal interaction,
it is demonstrated that speech muscle tension falls much more rapidly
following praise than following criticism. The same phenomenon was
noted in both the examiner and the patients studied. That is, after the
examiner had been critical his tension remained high in contrast to a
rapidly falling tension after he had praised.

Malmo et al. suggest that praise is similar to task completion
in its effect on muscle tension. In this connection they refer to another study from their own laboratory, a study by Smith (1953) showing
that muscle tension remains higher after task interruption than after
task completion. The term 'closure' (as used by Lewin to indicate the
discharge of a tension system) seems to have explanatory significance
in the area of muscular tension, they maintain.

That the concept of closure doesn't have any _general_ applicability
as an explanatory principle is brought out in a couple of case studies
also presented by Malmo et al. In one of these studies (1950) it is
pointed out that the patient showed considerable tension decrease during
"talking of extremely superficial matters which had been discussed
previously on several occasions". Another case study, by Davis and Malmo
(1951) illustrates clearly the very same point. In this study, focused
on a 27-year-old female, diagnosed as a severe anxiety-hysteric with
phobias, it was found that the patient's forehead tension was **higher**
in treatment hours in which the patient showed a constructive and
responsible approach to her problems than in hours in which her thinking
was confused, evasive and disorganized. In these latter interview hours,
the patient ascribed all her problems to a basis in organic illness, and
when important topics were introduced by the therapist, the patient
either broke off the discussion or expressed complete loss of memory of
what had been said earlier. In the productive hours the patient accepted
the fact that her difficulties were of a psychogenic nature and insights
gained during these interviews were meaningful to the patient and subsequently effected changes in her thinking and feelings, Davis and Malmo
maintain. And they continue:

"The finding of the marked difference between levels of tension
is surprising in that it is the reverse of what might have been
predicted from the patient's reaction. During the productive
interview the patient seemed calm, reasonable, quiet, forward-
looking and fairly sure of herself. In the nonproductive ones
her thinking was disorganized, she was hostile and confused,
and appeared irritable and disturbed. During the productive
interviews there was an association of specific content (i.e.
anxiety-producing content like fears of injuring herself and
others, fears of insanity, etc.) with tensional phenomena.
This was not seen in the nonproductive interviews." (p. 913).

Davis and Malmo do not offer any explanation of their findings
besides suggesting that greater tolerance for anxiety is associated
with increased muscular tension, that the patient's anxiety level was
greater on the days of the nonproductive interviews, the anxiety level
of the patient on these days being so high that avoidance mechanisms
had to be resorted to.

One may ask what was actually avoided in the latter case.
It can hardly have been anxiety. Behaviorally speaking, the patient
was much more anxious (confused, evasive, disorganized) when the tension
was low. Actually we may infere that anxiety was avoided when the
muscle tension was high, that increased muscle tension represents a
mechanism of binding and avoiding anxiety.

However, we may ask if anxiety is not also an avoidance
mechanism? We may say for instance, that the patient by her anxious
behavior avoided the establishment of contact, of taking responsibility
for herself and representing herself vis-a-vis the therapist and herself. As the reason for this contact avoidance we may infere a set of
false expectations on the part of the patient of what would happen if
contact was established, e.g. that it would arouse so much excitement
and tension that it would be overwhelming and unbearable. As noted,
the tension level increased when contact was established, but so did
also the patient's self reliance so that the tension-increase could
easily be dealt with in constructive manner.

Our reasoning above follows broadly the train of thought of
Perls et al. (1951). Anxiety, they maintain, is produced by suppression
of contact, it is excitement prevented from being lived through due to
an impulse-defense confluence. The establishment of contact increases
an individual's energy mobilization, but at the same time also his ability
to cope with the energy mobilized.

The prevention of contact can take various forms, Perls et al. maintain. It can take the form of verbalizing, intellectualizing, pseudo-objective theorizing, detachment, evasiveness, hyperemotionality, and histrionic behavior.

Does contact avoidance take place on a purely psychological level without observable somatic counterparts? Perls et al. don't think so. Contact avoidance decreases an individual's energy mobilization, that is, we should expect a decrease in muscular tension. We may think of this decrease as a hypotonicity response - or as an expression of tensional configurations being "drained off". In the latter instance, the reduction of tension in certain muscle groups should parallel a tension increase in other "focal" muscle groups. In the former instance, we are led to assume that hypotonicity may function as a "repressive agent" in relation to hypertonicity.

It has been pointed out (e.g. by Braatöy, 1954) that in the persistent suppression of affects and drives hyper- and hypotension may show a vicariousness in function.

The hypothesis that contact avoidance may reduce muscular tension bring us over to the question of latent and manifest tension. During contact avoidance we should expect tensional configuration to exist in a state of latency. It is interesting to recall Davis and Malmo's observations that an association between specific content and tensional responses was only found while the patient was "quiet, forward-looking and fairly sure of herself."

The question of latent and manifest tension.

Assuming the existence of latent muscular tension, we may start out by asking: What conditions are necessary in order to mobilize latent tension?

If manifest muscular tension is looked upon as an expression of a psychodynamic conflict and a necessary stage in the resolution of such conflicts, all psychotherapeutic as well as physiotherapeutic techniques claiming to resolve conflicts would have to be considered potentially as possessing relevant conditions. Limiting ourselves to an experimental framework two viewpoints have been suggested, the one stressing specific, the other, non-specific situational stimulation.

In an article by Malmo (1959) it is maintained - based upon results from experiments with animals - that an organism's tension or activation level may increase monotonically with the amount of deprivation of a biological need (e.g. thirst) but in such a way that the monotonically increase in tension will be observed only under appropriate environmental conditions; if these conditions are not present, the tension at any one time will be latent and unobservable, and continue to be so until the appropriate external stimulation is presented, that is, until a situation exists which is mean-goal related in one way or another to the need in question. From such a viewpoint we should expect only specific stimulating conditions to elicit latent postural tension. We should for instance expect only a psyciatric interview focused on sex or a sexual affair proper, to provoke pelvic and leg tension in a subject possessing sexual conflicts.

The viewpoint just mentioned contrasts somewhat with the opinion presented in some earlier works by the same author (Malmo, 1957). Here it is stated that nonspecific stimulation, specifically when it presents a certain component of stress, is the best agent to bring out significant individual differences in tension patterns. For instance, in discussing the difference between mental patients and controls, Malmo maintains that it is not necessary to present material or stimuli to which patients are specifically sensitized or to present them with situations which have specific associative meaning for them. It will always suffice, he maintains, with a nonspecific stress situation, e.g. a standard situation involving pain stimulation to the forehead or hand, or a strong auditory stimulation.

The two viewpoints mentioned may be interpreted as being in disagreement, but may also be considered as supplementing each other and pointing to somewhat different tensional phenomena. For instance, it might be that the manifestation of some tension patterns presupposes drive-related stimulation, while others are manifested always and everywhere the individual is confronted with nonspecific stress in one form or another. We may think of the latter type of tensional patterns as a sort of "higher order" defenses draining off excitation in order to maintain a sort of optimal homeostatic balance.

Some observations by Malmo (1959) are highly relevant to this very last point. In an investigation of the effect of sleep deprivation

on the activation level of three young male non-patient subjects, Malmo
found one muscle area, a different one for each subject, to show
significant rise in tension over the vigil. He states:

> "It was the neck muscle in one $\underline{S}$, the forehead in another and
> the biceps muscle of the right arm in the third. In each case
> the one muscle showed statistically significant rise in tension,
> and in none of the $\underline{S}$'s was there significant tensional rise in
> any other muscle. In fact, there was regularly progressive and
> very significant fall in the tension of the left forearm in all
> three $\underline{S}$'s....Where high level activation is long continued as in
> a vigil or in certain psychoneurotic patients, it appears that
> skeletal tension may become localized to a single muscle group....
> It should be noted that in one-session experiments, where rise in
> activation was for relatively short intervals of time, tensional
> rise occurred in more than one muscle group." (p. 383).

Malmo's observations, although not directly focused on 'contact
avoidance', suggests that such a mechanism, probably being initiated
and maintained as a homeostatic device, will be accompanied by localized
tension in specific muscle groups.

The question confronting us is which muscle groups are most
likely to be involved. Burrow (1953) as well as Braatöy (1947) have
expressed definite opinions on this question. Both imply that 'contact
avoidance' has specific somatic counterparts - specifically in breathing
and eye tension.

According to Burrow an individual can adapt to his environment
in terms of either affecto-symbolic reactions, reactions based upon
personalistic experiences mediated through symbols and their adherent
affects, or in terms of spontaneous organismic reactions, reactions
based upon an immediate contact (behavioral solidarity) between own body
and the surrounding world. In the treatment of psychiatric disorders,
what is really called for is not the analysis or solution of affecto-
symbolic conflicts, Burrow maintains, but an alteration of the level
of adaptation. This alteration may be brought about by techniques
focused on the regaining of kinesthetic control, specifically, kinestetic
control of the eyes since the eye muscles to a larger extent than any
other muscle groups play a focal role in the blocking of an organism's
immediate relation to itself and the environment.

Going back to the case studied by Davis and Malmo one may
hypothesize that the patient in her most unproductive hours, in the
hours where the frontalis muscle tension was low, exhibited the most
pronounced tension in certain other muscles, i.e. certain eye muscles.
It is commonplace observation that psychotic behavior disorders very
often express an 'empty' gaze. The peculiar gaze of psychotic patients
has been commented upon time and again, and that this gaze corresponds
to certain muscle tensional phenomena seems likely.

In his attempts to go beyond words, images and symbols, Burrow's
approach has some similarities to yoga conceptions. Of specific interest is the respiratory changes reported by Burrow, as concomitant
with the mental changes taking place through sustained attempts to gain
kinesthetic arrest and awareness of eye movements. When consistent
awareness of the tensions within and behind the eyes is obtained, Burrow
states, there is invariably a reduction in the rate of respiration. In
an experimental study, the reduction was found to be highly significant,
from an average of 13.22 respiration per minute in normal attention, to
an average of 4.43 in what Burrow calls cotention. The amount of oxygen
absorbed in the two states was found to be very much the same, the
greater oxygen utilization in the cotentive state probably resulting
from the deeper breathing found under this latter condition.

Burrow's data suggests that diminution of affecto-symbolic
images, diminution of eye movements and the slowing of respiration, belong to one consistent pattern. However, one may question Burrow's
assumption that this behavioral pattern is an organismic adaptational one.
In fact, the pattern may just as well be considered a defensive pattern,
a pattern characterized by withdrawal from "affects" and from immediate
contact and relation to the environment. Instead of releasing affects,
releasing and resolving bent-up conflictual impulses, the approach is
mainly directed toward opposite ends, the going beyond (repressing
still further?) interpersonal feelings, impulses, and affects.

We want to stress this last point specifically because just
here we find a pronounced difference between Burrow and Braatöy. Braatöy
too emphasizes the 'pathological' aspect of many people's intellectual
orientation, although he does not think that the affecto-symbolic level
of adaptation by necessity is an unhealthy or non-organismic one. The

question is primarily whether the affecto-symbolic level is immediate, self-represented, contact-loaded, or not, whether it is integrative or defensive in structure and function. If defensive, nothing is gained by going beyond it--it would have to be worked-through, dissolved and re-integrated, and in this process, a mobilization of the defended-against impulses would have to be accomplished in one way or another. It is, however, of interest to note that in order to accomplish such a mobilization, Braatöy time and again emphasizes the great significance of the release of a patient's restricted breathing.[1] The therapeutic approach for which Braatöy is a spokesman, is to liberate, release and resolve unconscious conflicts and affects through the working through of defenses, and in this respect, the most conspicuous is not the particular postural defense against a certain impulse pattern, but the individual's contact-avoidance or contact-defense--spesifically manifested in the eyes and in the respiratory muscles. From the point of view of therapy the aim would often be to increase tensional phenomena, to get them into observable manifestations.

Granted such a viewpoint, that is, granted that contact avoidance is mediated mainly through the respiratory muscles and the eye muscles we are faced with a sort of vicariousness of function. We may ask: Will an individual with unresolved conflicts always and everywhere show 'inappropriate' tensional manifestations?

----

[1] In this respec Braatöy is following Reich (1942). "If one asks the patients to breathe deeply," Reich states, "they usually force the air in and out in an artificial manner. This voluntary behavior serves only to prevent the natural vegetative rhythm of respiration. It is unmarked as an inhibition: the patient is asked to breathe without effort, that is, not to do breathing exercises, as he would like to do. After five to ten breaths, respiration usually becomes deeper and the first inhibitions make their appearance." (p. 297). In other words, deeper breathing is counteracted by other defenses, the increased depth of breathing functioning as a release mechanism for the mobilization of other postural patterns in order to maintain a given homeostatic balance. As examples of such patterns, Reich mentions the following: The head becomes stretched forward or it starts to jerk from one side to the other, the shoulders become tightly held or pulled up or back at the end of each expiration, the back becomes arched, the pelvis becomes retracted and the upper abdomen protruded, the mouth becomes tensely closed, the eyebrows become knit, the legs and feet become rigidly extended, the patient starts to talk compulsively, or to stare vacantly into a corner or out of the window. Thus we end up with the hypothesis that inhibited respiration may serve as a higher order mechanism of defense in relation to primary defenses manifested as hypertonicity in particular muscle groups.

This latter question brings us back to Malmo's experimental findings. In an article some years ago (1957) he states:

"(A) question which we sought to answer was whether a certain level of arousal must be reached in order to demonstrate differences between patients and non-patients or whether such differences could be obtained under resting, 'basal' conditions.... Our findings did indeed clearly show that, in differentiating between patients and controls, some form of stimulation was definitely superior to merely taking records under resting conditions. This has been demonstrated for blood pressure, for muscle potentials in motor tasks, and again for muscle potentials in two separate investigations of startle. The only measure which we have found to discriminate between patients and controls under 'resting' conditions was frontalis-muscle potentials. However, we know that 'resting' conditions associated with a testing session are by no means basal, and that--for example-- significantly lower blood pressure readings may be obtained from patients resting quietly on the ward than in the socalled 'resting' condition of our experiments." (p.280).

We fully agree with Malmo that resting conditions associated with a testing session are by no means basal. However, the question arises whether a basal condition for EMG recordings is anything more than a hypothetical base-point on an activation continuum, a base-point which has to be defined in terms of EMG responses. If so, we may very well be confronted with the fact that some individuals never show these base-point responses regardless of the degree of external basal conditions present. Consequently, to aim at basal conditions will easily lead to a rather fruitless circular form of reasoning.

If non-specific stress is sufficient to bring out in the open differences between mental patients and controls we may hypothesize that the very same differences are present, although less detectable and measurable, in a resting situation. This follows from the assumption that a resting situation for most subjects will not represent a basal but a slightly activating situation, that is, many subjects will always and everywhere exhibit a certain tensional activation and irrational preparedness.

We would like to reemphasize Malmo's finding that frontalis muscle tension discriminates between mental patients and controls under resting conditions. This result is of particular interest because of the physiognomic closeness between frontalis and eye muscles, although

Malmo et al. (1954) argue that their frontalis measure probably did
not tap eye muscle tension. Worth recalling is also Haavardsholm's
(1946) results previously referred to - suggesting significant differences under resting conditions between healthy, neurotic and
psychotic subjects in their I/E ratio of EMG-potentials from the
diaphragm muscle, and somewhat parallel to these results, Clausen's
(1951) findings as regards differences in breathing patterns in the
same nosological groups under resting conditions. Both of these latter
studies are in need of independent cross-validation, although the results
support each other and also fit well into the same theoretical
framework.[1]

Let us briefly try to sum up the tensional phenomena we have
been considering so far.

In the <u>first place</u> we have the fairly generalized tension increase accompanying attention, alertness, and a situationally induced
rise in an individual's activation level generally.

<u>Secondly</u>, we have the more specific and localized tension
increase accompanying the presence of a stimulus situation related to
a drive system or to a latent impulse-defense constellation.

<u>Thirdly</u>, we have the more specific and localized tension in-
crease found under non-specific stress stimulation and when a high level
of activation is continued for a relatively long time.

<u>Fourthly</u>, we have the tensional discharges characterizing body
movements.

----

[1] Both studies were done as dissertations at the Psychological Institute
at the University of Oslo. In mentioning the need for independent cross-
validation we reemphasize only the researchers' own precautions.
But we would also like to stress that much more refined methods
exist today (as compared to two decades ago) for the measurement
of integrated EMG potentials and respiratory movements. On the
other hand, it might be seriously doubted if the focusing on ill-
defined nosological groups is a wise procedure. Although we fully
acknowledge the very convincing findings of, for instance, several
of Malmo's experiments on patients and controls, we still think
that the linking of EMG and respiratory differences to behavioral
variables might in the long run be more profitable. Consequently
we would hope that future studies would concentrate on behavioral
variables (based upon rating scales on psychological tests) like
"ego strength", "excitement level", "maturity of ego organization",
"manifest anxiety", "field vs. body dependence", and variables
taking into account the point made by Jahoda and others, that
"mental sickness" and "positive mental health" are not necessarily
opposites in one and the same behavioral dimension.

Fifthly, we have the specific tensional changes accompanying what has been called conscious ideomotor activity.

Sixthly, we have the postural changes accompanying anti-gravidity shifts in body position.

Seventhly and finally, we have the more specific and more generalized tension patterns characterizing individuals under resting conditions.

In the last section we have paid specific attention to the last point mentioned. A number of questions emerge:What is the relation between habitual tension patterns under resting conditions and the patterns found under non-specific stress stimulation, and when a high level of activation is continued for a relatively long time? What is the relation between habitual tension patterns under rest and the localized tension increase sometimes found in situations related to specific drive systems? What is the decisive factors shaping an individuals habitual resting pattern ?

In previous sections we have given some tentative answers to a couple of these questions. We have suggested that the localized tension increase found under drive-specific stimulating conditions is dependent upon habitual tension patterns, and that the same probably is true as regards tension patterns found under non-specific stress. However, the relationships involved is very complex. The main reason for this complexity is due to the fact that an individual's habitual tension patterns is only one facet of his characterological make-up in terms of tensional configurations. As in the realm of psychology proper we may talk about displacements and manifest patterns hiding latent ones. It is true that by concentrating on postures we are going behind words and beyond an individual's conscious self-expressions, but we are still confronted with the question of manifest content con-cealing its underlying roots, the question of what is behind what is behind. In approaching this question we are not in a position to cite many experimental findings or to refer to laboratory investigations. We are moving into an area in the outskirts of experimental psychology, an area, which so far has been mainly the concern of clinicians.

What determines tension patterns under resting conditions?

In his article "Analysis of postural behavior", Deutsch (1947) states:

"When a patient is invited to lie on the couch and relax, his posture illustrates not muscular relaxation but a pattern of behavior related to the situation and to a basic psychosomatic pattern." (p. 195).

By basic psychosomatic pattern Deutsch refers to the postural pattern which is characteristic of an individual and expresses his basic psychosomatic homeostasis. In order to demonstrate this point Deutsch presents a number of his personal observations, how certain habitual postures seem to change parallel to changes taking place in an individual's psychic structure.

Going into the area of the relationship between postural patterns and psychic structure generally, the most comprehensive theoretical model existing today is probably the one launched by Reich in the middle thirties in the short period between his psychoanalytic train of thought, giving rise to his character analytic technique, and his total involvement in bionic and orgonomic speculations.

Reich's basic assumption is very much the same one as Deutsch's, namely that a certain psychic structure is at the same time a certain biophysiological structure, and that the experiential world of the past is alive in the present in the form of character attitudes, attitudes expressed simultaneously in terms of postural patterns (muscular attitudes) and modes of behavior.

A central aspect in Reich's reasoning is the concept of "layering" or "stratification of defenses." This layering is assumed to take place both in the psychic and in the somatic realm. Describing the interlacing of defensive forces from a psychological point of view, he gives the following illustration (1949):

"If, for example, one has unmasked an over-polite attitude which forms the top layer, as a defense function, that which was warded off makes its appearance, say, aggression. It would be wrong at this point to tell the patient that he is living out his infantile aggression, even if this aggression appears in an unmistakable manner. This aggression is not only the expression of an infantile

relationship toward the world, but at the same time itself a
defense against something which is more deep-lying, against,
say, passive anal impulses. If one succeeds in eliminating this
defense layer also, it may happen that what appears is not the
expected passivity, but contactlessness, in the form of in-
difference toward the analyst, etc. This contactlessness is
unequivocally a defense, say against an anticipated disappointment. If one succeeds, by dissolving the contactlessness, in
bringing the fear of disappointment to the surface, it may have
all the appearances of a deep infantile fear of losing the love
object, but at the same time it is defense against deep aggressive impulses towards the love object which once withdrew its
love. This example could be varied, complicated or simplified
indefinitely, according to the type concerned." (p. 315).

Reich's point is that layers of defenses are built up on the
principle that impulses being warded off very often serve the function
of warding off still more deeply repressed impulses. In another connection, describing his therapeutic experiences, he states (1942):

"The neurosis of each individual patient revealed a specific
structure. The structure of the neurosis corresponded to the
development. That which had been repressed latest in childhood
was found to lie nearest the surface. However, early infantile
fixations, if they covered later conflicts, could be dynamically
deep and superficial. For example, the oral fixation of a woman
to her husband, deriving from a deep fixation to her mother's
breast, may belong to the most superficial layer of character
when she has to ward off genital anxiety toward her husband...
As a rule, the structure of the neurosis corresponds to the
development, but in reverse order...The layers in the character
may be compared to geological or archeological strata which,
similarly, are solidified history...Each of these layers in the
character structure is a piece of life history which is preserved
in another form and is still active. It was shown that by
loosening up these layers, the old conflicts would--more or less
easily--be revived." (p.122).

Since Reich equates character attitudes with muscular attitudes, it follows that also within the latter realm one would expect
to find stratifications, a layering of muscular tensions from manifest
superficial ones to latent ones. Consequently, when invited to lie
on the couch and relax, an individual will exhibit the top layer of
his muscular attitudes. As stated, this top layer will usually
correspond to the individual's latest defense-synthesizing endeavors,
but may also include the direct expression of early infantile fix-

ations if these are specifically activated by the individual's present life situation.[1]

In a case discussion, Reich (1942) attempts to illustrate how various facial expressions may be stratified and how this stratification may be revealed through analytic treatment:

"From the very start of the treatment, the 'indifferent' expression of her face was striking...When one would talk to her, even about serious subjects, she would always stare into a corner of the room or out of the window. With this, her face would wear an indifferent expression and her eyes would have an empty, 'lost' look. As this indifferent expression was thoroughly analyzed and dissolved, a different expression appeared... The region of the mouth and chin were 'angry', while eyes and forehead were 'dead'...I proceeded first to work out separately the expression in mouth and chin, in the course of this work there developed incredibly violent reactions of inhibited impulses to bite...After about two weeks' work at the mouth region, the angry expression disappeared completely in connection with the analysis of a very intense reaction of disappointment. After the attitude of mouth and chin were dissolved....the expression of eyes and forehead gradually became...observing, critical and attentive...(And) to the same extent to which the 'dead' expression was replaced by the !critical' expression, the defense against genitality became accentuated. Following this, the critical, severe expression began to alternate with a cheerful, somewhat childlike expression in forehead and eyes...With the final disappearance of the critical attitude of the forehead and its replacement by the cheerful attitude, the inhibition of genital excitation disappeared also." (p. 317).

As in the psychic so also in the somatic realm we seem to be faced with mechanisms of displacement. Commenting upon this latter phenomenon, Reich (1949) states:

"The mechanisms of the pathological displacements and fixations of the vegetative energies may be hidden in such phenomena as the following: a weak voice which hardly sounds at all, a lack of movement of the mouth in talking, a slight mask-like facial expression, a slight indication of the facial expression of a suckling infant, an unobtrusive wrinkling of the forehead, drooping eyelids, a tension in the scalp....a certain way of holding the head to one side, of shaking it, etc. One will find that the fear of genital contact does not make its appearance as long as these systems in the head and neck regions have not been uncovered and eliminated. Genital anxiety, in most cases, is displaced from below upwards and is bound in the hypertonus of the musculature of the neck." (p.345).

- - - -

[1] Of course we are here confronted with the type of situational impact referred to by psychoanalysts as the transference phenomenon.

In order to dissolve genital conflicts basically expressed
in a chronic tightening of the pelvis and the legs, it would be use-
less, Reich maintains, to try to dissolve the pelvic tension right
at the beginning. Usually, the dissolution of genital conflicts will
have to begin at places that are far away from the genital apparatus,
mostly at the head. The facial expression and the character of the
voice are frequently those functions which an individual is most likely
to be aware of, and parallel to that, also those manifestations which
represent the most superficial layer of defense. If, for instance,
one has dissolved tension in the forehead, a stiffness in the neck,
or a spasm in the throat or the chin, there almost regularily appears
some kind of impulses in the chest and the shoulders, but before long,
these are curbed by corresponding inhibitions. If these inhibitions
are dissolved, impulses in the abdomen may appear until these too meet
an inhibition, and not until these inhibitions have also been dissolved
will it be possible to start releasing pelvis tension so that the
patient may regain his natural pelvic motility.

Such a description should of course not be taken too literally,
i.e. individual differences in developmental history may complicate and
modify the picture. Although Reich has a tendency to _genitalize_ all
sorts of conflicts, we may still consider his model as somewhat content-
free in nature. The dynamic root of the character structure may not
involve genital conflicts, but oral ones with a repressive nucleus in
the mouth, the neck and the throat region, or anal ones, with a
repressive nucleus in an initial holding back of feces by the tensoning
of the buttocks. However, the main principle might be the same, that
only through the mobilization of certain impulses will inhibitions that
have previously remained hidden become unmistakably evident, i.e. the
heightened excitability of the warded off impulses mobilizing further,
so to speak, the latent inhibitory mechanisms.

From what has just been said it follows that what is expressed
overtly and posturally by an individual lying in a "relaxed" position
is only a fragment of his personality structure and frequently a most
superficial fragment. To get at deeper levels of the personality
structure it would be necessary to mobilize various impulse patterns
and study the individual's reactions to such an induced heightened

excitability. One way by which such a mobilization may take place is through the discussion of certain topics in a therapeutically oriented interview (cf. Shagass and Malmo's (1954) studies), another way, through the presentation of visual or cinematographic stimuli pertaining to specific impulses (cf. Sines, 1957), and finally a third way, through passive movements of the muscle groups involved in the execution of various impulse patterns, i.e. through recording of the resistance mobilized in the musculature of the jaws when the chin is moved downward and upward, through observation of the resistance mobilized in the legs, when the legs are adducted and abducted, flexed and streched, etc.; the passive movement in these instances forcing the individual into postural patterns representing the substratum for the triggering off of various basic responses. Since the repression of these responses would be jeopardized by the presence of the appropriate postures particular passive movements will immediately trigger off defensive counter measures in the form of active resistances (hypertonicity), lifeless flabbishness (hypotonicity), or simulated participation.[1]

Following Braatöy (1947) and Grieg et al. (1957) one may differentiate between various degrees of muscular resistances as well as between resistances of different types. As regards simulated participation some explanation may perhaps be in order. We are here referring to observations by Reich and others, that affective, spontaneous, infantile impulses (movements) may very often be warded off by simulated substitute movements, movements of an "acquired" more or less voluntary-programatic nature, with the same direction as the warded off ones. Voluntary movements of certain muscle groups may serve as a defence against a spontaneous impulse release. Consequently, in such a case it would not be possible to elicit basic responses before the substitute responses are unmasked and eliminated. Diagnostically speaking, a substitute movement, i.e. a simulated participation in passive movement, may be just as conflict revealing as a pronounced hypertonicity or hypotonicity.

----

1) A systematic diagnostic approach based upon this latter type of procedure is represented by Waal's method for somatic psychodiagnostics. (cf. Grieg et al., 1957). This is a standardized series of postural investigations - covering altogether 112 variables. It is of interest to note that from this method, valid or not, there emerges a dynamic and structural psychological personality description quite similar to one derived from an extensive battery of psychological tests.

So far we have been concerned mainly with hypertensions, although hypotension may be considered of no less importance from the point of view of impulse inhibition.[1] Hypotonicity has been much less systematically discussed in the literature. By and large it has been referred to, mostly in passing as a characteristic feature of certain personality types. For instance, Reich (1949) states:

> "While the typical compulsive character develops a general muscular rigidity, we find in other patients a rigidity in certain regions combined with a flaccidity (hypotonus) in other regions. This is particularly frequent in passive-feminine characters." (p. 344).

Lowen (1957) too, in his discussion of passive-feminine characters points out that their muscles often are soft or flabbish. The same picture may be present in schizoid characters, Lowen maintains. In both cases, however, the hypotonicity is limited to the superficial muscles. The deep muscles, e.g. the diaphragm muscle, the deep abdominal muscles, and the muscles of the pelvic floor, are practically always spastic. Thus, the hypotonicity of superficial muscles is considered an expression of fairly deep inhibitions, specifically of aggressive and self-assertive drives, inhibitions which have to be transferred into hypertonicity in order to be therapeutically accessible.

In the area of relaxation therapy (e.g. J.H. Schultz and E. Jacobson), much more attention has been paid to hyper- than to hypotension. While the detrimental effects of high tension has been greatly emphasized, little or nothing has been said about the psychological aspects of muscular flabbishness and hypotension. In the area of psychophysiology too, so far much more work has been devoted to the study of hypertonicity. A particular reason for this has been the lack of adequate recording equipment. For instance, among EMG specialists

----

1) A specific case is the rapid shifts and inconsistencies sometimes observed in psychiatric patients. Describing a case study of an ego-weak 27-year-old housewife, diagnosed as a severe anxiety-hysteric with phobias, Davis and Malmo (1951) state: "(The patient) personality type was impulsive, variable, and dependent, and she had a multiplicity of symptoms usually of short duration and constantly changing pattern. This was associated with rapidly variable levels of tension in all areas recorded from, and no persistent high levels in any area."(p. 914). It is probable that the tensional pattern described is characteristic of impulsive, ego-weak personalities generally.

the absence of measurable EMG's often found during relaxation has been
interpreted partly as an indication of tonic muscles being different
from phasic ones and being innervated by impulses unmeasurable by
EMG equipment, partly as a lack of electrical activity altogether.
It is our strong conviction that increased knowledge about hypotonicity
would enhance our understanding of hypertonicity as well, in fact,
that further insights into muscular tonicity generally to some extent
are dependent upon a "break through" in the former area.[1]

       We started this latter section by asking what determines
tension patterns under resting conditions. Our discussion has indicated
that central personality variables (in terms of unresolved conflicts
and impulses) might be involved in maintaining tension everywhere and
always, although the principle of "an optimum homeostatic balance"
may displace tension into individually distinctive patterns far removed
from the muscle groups specifically involved in the origin of the tension
in question. How these principles operate we are still unable to say.
We might hypothesize that a specific type of indentification process
is of importance, although this would mostly be a question of words as
long as we don't know the nature of the processes in question. We might
hypothesize, further more, that underneath spesific infantile impulse-
defense configurations are involved. In the last chapter we tried to
outline a developmental model concerning these configurations. Although our review of the experimental litterature pertaining to postural
dynamics have brought out some relevant and "confirming" empirical findings, the model is mainly pointing to phenomena that have so far been
outside the realm of experimental investigations. In the last section
we have in a sense been confronted with a cross-sectional perspective of
the very same phenomena that is dealt with from a longitudinal point of
view in the last chapter.

----

1) Probably the first step would be the introduction of more sensitive
   instruments than the usual EMG techniques. A method offering some
   hope is the measurement of muscular tonus through the recordings of
   muscular microvibrations. The phenomenon of muscular microvibrations
   has been extensively discussed by Dr. Hubert Rohracher at the
   Psychological Institute of the University of Vienna. Microvibrations
   refer to microscopically small, ever ongoing movements or contractions
   of the human body. Their frequencies are between 7 and 11 per second
   and their amplitude between 1/1000 and 5/1000 of a millimeter in the
   relaxed and resting subject. On the basis of his investigations
   Rohracher suggests that the rate or frequency of the vibrations corresponds to the amount of heat being produced by chemical processes in
   the muscles, and that the size of their amplitudes corresponds the
   level of psychophysical or tonic tension characterizing the individual.

POSTURES AND PERCEPTUAL BEHAVIOR.

In reviewing Deutsch's case histories in the first chapter
we noted time and again that postural changes preceded or accompanied
changes in the patient's verbalized material. We noted statements
like: "He relaxed his hands and remembered acts of rebellion,"
"...increased muscular tonus was a forerunner for verbal expressions
of self-assertiveness and hostility", "...in verbalizing these thoughts
he sprawled his legs," etc. In other words, specific postural configurations seemed to accompany and covary with the content of
spontaneous ideational expressions. With this observation in mind let
us turn to the area of perceptual behavior.

Following Bruner's (1957) definition of perception as a process
of categorization in which organisms move inferentially from cues to
categorical identities representatial in nature, i.e. with varying
degrees of veridicality; and his definition of perceptual readiness as
the accessibility of certain categories for use in coding or identifying
environmental events, we may consider postures as the somatic counterpart of perceptual readiness. Postures represent readiness to encounter
certain objects and events. A person in a boxing position, with clenched
fists and raised hands, is ready to encounter attacks and threats; a
person with arms stretched out and rotated backward, ready to encounter
and embrace a loved object; a female lying with spread out legs, ready
to encounter a male sexual partner (at least more so than if her legs
are rigidly crossed); a person with hands clasped over his chest, ready
to encounter a nursing, protective and nurturing mother, etc., etc.
We would expect these readiness-to-encounter-postures to have perceptual
consequences, i.e. that they represent readiness for perceptual as well
as for overt behavioral responses. In other words, we think it is
reasonable to conceive of postures as representing "categories" of high
accessibility.

Following our earlier train of thought about postural defenses,
the fact that certain responses may be inhibited by a dissolution of

their postural basis, we end up with the proposition that such a dissolution will have repercussions too on the perceptual accessibility of the response-categories in question. In other words, at the same time as a certain posture represents an increased accessibility of specific responses so does it represent a decreased accessibility of other responses.

In his book "Theories of Perception and the Concept of Structure" Allport (1955) points out that motor aspects of perception so far has been a quite neglected area in the psychology of perception. He deals at length with Werner and Wapner's sensory-tonic field theory and their experimental demonstrations that tonic and proprioceptive factors do have effect upon the perception of objects and events. Of special interest from a more clinical point of view is their finding that following a situation in which motor activities are inhibited, observers tend to make more "movement" interpretations of a Rorschach card than when under uninhibited conditions. This fits of course well with the common assumption that lack of M responses is characteristic of infantile "impulse-ridden" characters.

Discussing the phenomenon of set, Allport writes:

"We do not usually perceive "out of the blue". Unless we are startled or confronted with an unexpected situation there is behind our perceptions, as we go about our daily affairs, a background state of the bodily musculature that is more or less relevant to the objects we are about to perceive." (p. 210).

Going further into these processes, Allport states:

"The entire tension-pattern at any given time consists of two parts. First, there is a general, or background, supporting function consisting of diffuse tensions in muscles other than those concerned with the particular activity. Second, there is the specific supporting functions consisting of tensions in differentially localized muscle groups. This latter tension-region represents the immediate preparation and support for the ensuring or continuing activity for which the organism is set. It serves as the "focus" of the tension-pattern; and the focus of the preparatory tension-set tends, on experimental evidence, to be identical with the muscles whose tensions are involved in the full overt response. The general tensions in other muscles serve as background excitation and support, both peripherally and centrally, of the focal set. There is 'thus a gradiant of activity from the focalized tension-region outward to the other bodily musculature. General tension-effects from the background

excitation maintain the body in a state of tonicity during
waking activity and provide a flow of returning proprioceptive
stimulation constituting the "vigilance" of the central nervous
system. Body-tonus is maintained in a self-completing, circular-
reflex manner." (pp. 221-222).

Allport goes on stating that long-standing sets, peculiar
to the individual, may determine the characteristic content of his
perception, and calls attention to the possibility that the postural
substrate may become abnormally "fixated", especially in cases of a
persistent thwarting of motivations or needs, and that "backlash"
from postural tensions thus aroused may dominate the afferent inlets
so that it would be difficult for exteroceptive stimulation to break
through and produce shifts in behavior. Such postural fixations might
help to explain the loss of perceptual "contact with reality" as seen
in psychotic patients, and the same explanation may also be employed
to account for "autistic" and non-veridical perception, Allport maintains.
At this point it is interesting to note that laboratory experiments
have found a profound inability to maintain a response set a most
prominent behavioral trait among schizophrenic subjects (Shakow 1962).

Allport's hypothesis that muscular contractions, tonic states,
and proprioceptive backlash are universal factors in perception and
play an important part in determining how the world appears to the
perceiver, fits in very well with our own conceptions, although when
we talked about various postures increasing and decreasing the accessibility of specific perceptual responses we were mainly thinking about
interpersonal perception. Underlying our assumption was the conception
of person perception being basically empathetic in nature.

Person perception and empathy.

Empathy may be looked upon as a particular form of perception.
To emphathize is to diagnose, to recognize and identify feelings and
emotions in others by putting oneself in their position, and this
putting-oneself-in-their position implies, in our view, to imitate and
implicitely to duplicate postural configurations in others. It is in
a sense a non-intellectual form of perception although it is certainly
an inferential categorization process. We would assume empathy to be
a most basic form of person perception, that is, that in primitive organ-

isms perception and awareness and motoric responses are to some extent
the same act, that initially person perception is touching and imitation, and that also later on a close connection exists between person
perception and spontaneous implicit postural adjustments and accomoda-
tions.

In the case of a dissolution of the postural basis of certain
response categories we would expect an individual's perception to be
distorted when confronted with stimuli in his environment corresponding
to the response categories. Not being able to empathize (i.e.,
perceive by imitation) he will be restricted to deal with existing cues
in a completely intellectual manner, to relate to human beings as if
they were thing-objects. This being the case, veridical perception
will hinge upon how well his expectations correspond to environmental
events and how able he is to get a "constant close look."

Our conception of imitation as a basic form of person perception
corresponds closely to McDougall's (1920) theory of a direct induction
of feelings by means of primitive sympathetic reactions, although we
are inclined to agree with Allport's (1924) objection to the theory,
namely that a certain psychological predisposition has to be present
in order for a certain induction to occur.

Our differentiation between empathetic and intellectual perception corresponds to some extent to Murray's (1933) discussion of
the relationship between conscious and unconscious person perception.
When one looks at the face of another human being there seems to be
some retinal stimulation which is not translated into conscious perception, but which - through other effects - determines apperception,
Murry maintains. Certain physical features of the object which are
not consciously perceived, nevertheless affect the subject's body, and
even though the subject may not be able to report upon these internal
happenings, they may strongly affect his conscious appraisal of the
object. Murray suggests that what is involved may be a sort of unconscious imitation, activating feelings, emotions and kinesthetic
sensations, which in turn color the subject's conscious perception and
appraisal. He further suggests that learning might be of importance,
and leaves open the possibility that an unlearned physiognomic
responsiveness may also be involved in the process.

Keeping to our own theoretical framework we would assume a complete lack of empathy to occur relatively seldom. On the other hand, we would also think a complete lack of intellectualization to be relatively infrequent. Most often we would probably be faced with an intermixture of both types of processes.

In assuming that a complete lack of empathy will occur relatively seldom we don't thereby imply that person perception is characterized by a very high degree of veridicality. We would assume that empathy very often will take place spontaneously but that the emerging implicit posture, if threatening, will be instantly met by defensive counter measures of a postural nature resulting in autonomic arousal, perceptual blocking, or a perceptual categorization in terms of the most relevant category within the categorical limitation set by the subject's postural defense reactions. Of course, the possibility also exists that the most accessible category will be triggered off immediately without going the way through a 'veridical' imitation process. However, it would be difficult in this latter case to explain the autonomic arousal found in Sines' (1957) study when subjects were confronted with pictures depicting their main area of conflict, and perhaps also the results of McGinnies' (1949) and Lazarus and McCleary's (1951) studies, where autonomic responses were found to follow the presentation of a potentially traumatic stimulus without the subject being able to give a report of the nature of the stimulus in question. One may, of course, object that autonomic arousal is no proof of a precedent "subliminal" postural reaction.[1]

That involuntary muscular responses do get evoked by conflict-loaded visual stimuli is demonstrated in an investigation by Smock and Thompson (1954). Comparing verbal responses to a sample of Blacky Pictures with reaction time and involuntary muscular responses to selected stimulus-words (corresponding to the dimensions focused upon by the Blacky Pictures), the investigators found a significant positive correlation, that is, certain content categories (perceptual categories) in the first instance were found related to certain somatic responses (called anxiety reactions) in the second instance. Again, it may be

----

[1] In an earlier monograph (Christiansen, 1961) we have discussed at some length the present viewpoint as it relates to parent-child interaction, and to a subject's perceptual responses to the Blacky Pictures.

maintained that the results did not prove that muscular reactions
preceded perceptual ones.

The question of pre-perceptual reactions brings us over to
some investigations by Blum, these investigations, too, using Blacky
Pictures as stimulus material. In one of his studies presenting
Blacky Pictures tachistoscopically below threshold for conscious
awareness, Blum (1954) found that the picture the subjects felt "stood
out the most" was the very picture which probably depicted their main
area of conflict. In this as well as in another study (1955), Blum
further claims to have demonstrated that subjects, although specifically sensitized to conflict themes, will avoid "naming" or "guessing
the pressence of" pictures depicting these themes when being subliminal-
ly confronted with them. Both of these studies strongly suggest that
some form of "subliminal" reaction takes place to "subliminal" stimulation.
The question is, is this subliminal reaction basically of a postural
nature? Is the fact that veridical perception of conflict stimuli was
found to be significantly lower than chance an indication of something
"active" interfering between an instant preperceptual reaction and the
perceptual reaction proper? If so, is it possible to identify the
preperceptual reaction as well as the interfering mechanism within the
realm of the muscular system?

Very little research has been done in this area. Most relevant
is the work of Sokolow and other Russian psychophysiologists, although
their main interest has been focused on pre-perceptual responses within
the autonomic nervous system.

Traditionally, the autonomic nervous system has been regarded
as exclusively effector in its action. Recent findings indicate that
autonomic responses have effects analogous to those of sensory and motor
responses and that these effects is far from restricted to the topics
of emotion and stress. Russian researches, according to Razran (1961),
have found evidence for the presence of interoceptors not only in all
the main portions of the vascular and digestive systems, but also in
the lymphatic nodes, kidneys, spleen, lungs, salivary glands, and so on.
What is of specific interest in the present context, is the Russian
experiments on what has been called the orienting reflex. An integral
part of the reflex is considered vasomotor in nature. The reflex itself is described as an unconscious preparatory and preadaptive re-

action, a general vigilance response to novel stimuli - adapting the
organism for perception and action. Although much more work is needed
in this field before definite conclusions can be drawn, we are here
confronted with an area of research that might one day throw important
light on the problem of "perceptual defense" and also on the eventual
involvement of the skeletal muscular system.

Our assumption that person perception to a large extent is
governed by implicit imitation processes brings us over to a field of
study which most appropriately may be called interpersonal physiology.

In a recent study in this area (Coleman, et al., 1956), focused
on the relationship between affective states in a patient undergoing
psychotherapy and autonomic (cardiac) reactions of the therapist, it was
found that "...the therapist reacted physiologically to the patient's
emotional expression in a manner similar to the patient's physiological
reaction to his own emotions". In another case study from the same
laboratory (Di Mascio, et al., 1957), it is concluded that there is
"a tendency for the therapist's heart rate and lability to follow a
pattern similar to that of the patient", and furthermore, that there is
"a positiv 'physiological identification' of therapist with patient when
the latter is expressing 'tension' or 'tension release' but a negative
one when he is expressing direct aggression'." In a third study by
Malmo, et al. (1957), on the physiological reactions of an interviewer
and the one being interviewed, the presence of a close agreement in
responses is again pointed out. The results of this study - previously
referred to - revealed different physiological reactions in the patients
being interviewed, related to whether they were praised or criticized.
After being praised their speech-muscle tension fell rapidly, in contrast to a non-falling tension following criticism. A most interesting
aspect of the study, however, is that just the same reaction pattern
was found in the interviewer. Here too differential reactions were
noted, and here too tension remained high following the interviewer's
criticism of the patient in contrast to a rapidly falling tension following praise and support of the patient's behavior.

None of the studies mentioned above has been focused on
perceptual processes proper. They demonstrate, however, the possibility
of studying postural correlates of person perception in an "experimental"
setting. The correspondence noted between the therapist's and the

patient's physiological reactions, suggests that veridical person
perception might be a question of the accessibility of all sorts of
physiological responses in the perceiver so that autonomous imitation
processes may take place without being interfered with by the triggering
off of defensive mechanisms.

We may ask what sort of cues are of particular importance in
person perception. Without going into any extensive discussion of this
problem we would like to mention, that according to several scholars,
non-verbal ones are far more important than verbal ones. And among
non-verbal cues, both olfactory and respiratory ones have been specifically emphasized. Discussing the importance of respiratory cues,
Braatöy (1954) states:

> "...breathing - like emotions belongs to social biology. The
> observer immediately 'feels' whether a patient breathes freely
> or not....If you are uncomfortable in the presence of another
> person without being able to understand why, try to observe him
> and yourself from this respirational point of view. You can then
> sometimes discover that the discomfort-inducing person 'does not
> breathe' or breathes in a very restricted way. The haughty Lord
> Haw-Haw's way of talking is provoking not only because he evident-
> ly looks down on you, but also because in his lack of breathing
> through the nose he gives you directly the feeling that he thinks
> you stink. The discomfort induced by such persons is, however,
> even more basic. The over-distinguished; that is, over-controlled,
> emotion-repressing speaker restricts by his breathing your
> respiration." (p. 176)

That auditory respiratory stimuli may exert emotional and
cognitive changes too, are brought out in Kubie's (1953) description
of one of his experiments on hypnagogic reveries:

> "....when we placed microphones on the neck near the trachea
> and led the breath sounds through an amplifier to the patients'
> own ears, the amplified sound of the respiratory rhythm often
> exercised a strikingly hypnagogic effect. At times the patient
> went to sleep. Sometimes we did." (p. 35).

The two examples just presented are from the area of respiration.
We have chosen our examples purposely from this area since a previous
chapter was devoted entirely to an extensive discussion of respiration
as a psychological and not as a communicative phenomenon. Obviously this
latter aspect is just as important from the point of view of interpersonal perception.

The problem of perceptual self-correction.

Granted that drive states whether conflictual or not, will
express themselves in the postural system - resulting in an increased
accessibility of certain response categories, we are lead to assume
that drive states generally, if sufficiently intense, will have a tendency to facilitate non-veridical perception. One may ask: What about
the possibility of making the necessary internal corrections (to ensure
veridical perception) by means of an introspective analysis focused
upon bodily attitudes, postures and muscular changes? Introspective
analysis of postural and set-experiences represented an important area
of study for Titchner and his students at the turn of the century. In
principle - introspection of muscular changes as "conscious" happenings
should be possible because of the return flow to cortical centers of
afferent impulses from proprioceptors in the muscles, tendons and joints.
Our point is - that the possibility might be present in some instances
and not in others, depending upon whether defensive postural mechanisms
are brought into operation or not. Let us explain a little further
what we are thinking about.

In the case of a 'pure' drive state, we would neither expect
anxiety arousal nor tendencies to avoidance when 'basic-drive-relevant'
stimuli are presented. Likewise, in this case we would assume that
perceptual self-corrections could come into operation, the subject if
trained, being able to use the proprioceptive feedback from his tensional
state to correct his sense impressions. In the case of defensive
postures this would be much more difficult, not so much in relation to
the prevailing defense stratagem as to the response category being
warded off.

In order to exemplify this proposition let us take a young man
with strong castration anxiety (strong anxiety related to genital impulses), but the anxiety being warded off fairly efficiently by a
compensatory overmasculine defense stratagem disposing the man to
perceive all women as basically subversive and submissive thereby
justifying his own disregard for their feelings, their rights to self-
decision and self-expansion. Confronting this man with his "non-
veridical" perception of women-- eventually teaching him to observe and
sense the adductor tension in his legs and his thoracic immobilization

when together with women, or talking about women, will not necessarily provide "feed-back" for veridical perception. He might become more cautious but his perception will probably all the same most often turn in a non-veridical direction, i.e. more likely he will move inferentially from cues to the categorical antithesis of his former category-preference than to the response category he is characterologically and posturologically warding off in himself. Concretely, our young man might, for instance, start to perceive women as basically more honest, stronger and more self-confident than men, or the self-scrutinizing procedure may develop into an intellectualization detrimental for the young man's immediate affective contact with women as well as with his own bodily processes. Instead of spontaniety and 'warmth' his whole orientation may increasingly become more and more programatical. To the extent the warded off category is called for (in order to make his perception veridical) he might be just as far off as before the training process (the feed-back sensitization) started.

The above comments fit the observation made by many psycho-analysts that antitheses of responses often seem to occur close to-gether and relatively easily may replace each other. The nearness of opposites as a principle in personality organization has been stressed in several studies by Frenkel-Brunswik (e.g. 1951) amont others.

One may wonder if it wouldn't be possible to pursue the training process one step further, to sensitize the subject not only to hypertensions but also to hypotensions and to intellectualizing devices. It is our position that this would be practically impossible without resolving the underlying conflict. This follows from the assumption that the underlying conflict is maintained just because the subject has lost contact with the muscular expressions of the forces and counterforces involved. If by means of one technique or another we should manage to reestablish this contact, the conflict would probably resolve itself due to reintegrative psycho-biological processes. This viewpoint is very much in line with the one proposed by Perls, Hefferline and Goodman (1951). They give the following illustration:

"Suppose one inhibits sobbing by deliberate contraction of the diaphragm and this becomes habitual and unaware. Then the organism loses both activities—that is, the man who manipulates his functioning in this way can neither sob nor breathe freely.

Unable to sob, he never releases and gets finished with his
sadness; he cannot even properly remember what loss he is sad
about. The tendency to sob and the contraction of the diaphragm
against sobbing form a single stabilized battle-line of activity
and counter-activity, and this perpetuated warfare is isolated
from the rest of the personality.

The task of psychotherapy, obviously, is to bring back the
boundary of demarcation--the awareness of the parts, and, in
the specific instance, of the parts as consisting of crying and
diaphragm-contraction. Crying is a genuine need of a human
organism which has sustained loss. Aggression against crying--
in this instance, tightening of the diaphragm--has become a need
only through the establishment of confluence with 'the authorities'
who say 'big boys don't cry'. To dissolve the inhibition requires
that the confluent, bound energy of the opposed parts be again
differentiated into sobbing and aggression-against sobbing, that
the conflict be revived under present, more favorable circumstances, and that it be resolved. The resolution must include
not just one but both sides of the conflict. The sadness will be
released by crying it out once and for all. The aggression-
against-sobbing contrary to one's own natural functioning--will
be redirected outward against antibiological 'authorities'."
(p. 120).

Summarizing Perls et al., and our own position, we would
suggest that sensitivity training to proprioceptive feedback would be
unnecessary or of limited help in instances in which unresolved conflicts don't exist, and where they do exist, irrelevant as far as
perceptual veridicality is concerned if not extended to the point of
conflict resolution, i.e. the breaking up of the confluence between
impulses and defenses and bringing the parts back to awareness as
separate entities. Needless to say, this would provide new categories
for the subject's perceptual behavior.

The above comments brings us directly over to the relationship
between body image, body awareness and postural tension - a topic we are
going to focus upon in the next chapter. Before leaving the topic of
perception proper we would like to emphasize that a motor theory of
perception obviously only cover a certain aspect of perception. Our
discussion of this aspect has been rather discursive - aiming at two
things primarily: 1) offering some references to relevant litterature
in the field, and 2) pointing out some tentative hypotheses derived
from our conception of postural dynamics generally.

POSTURAL TENSION AND BODY AWARENESS.

Compared to exteroception, interoception is largely unconscious perception, although it may be considered the human organism's
main window toward its inner world. The term "body image" refers to
how an individual conceptualizes his private body domain, how he
organizes his perception of his body, and how he feels about himself
as a separate entity. Gerstmann (1958) offers the following comprehensive definition:

> "By body image, or body schema, is understood the inner picture
> or model which one forms in one's mind, of one's body or one's
> physical self, in the course of life; and which one carries with
> one unwittingly, that is, outside of central consciousness. It
> is a kind of inner mental diagram representing one's body as a
> whole, as well as its single parts and territories according to
> their location, shape, size, structural and functional differen-
> tiation and spatial interrelation. It also represents the car-
> dinal directions of the body; right and left, anterior and
> posterior, up and down. The body schema can thus be conceived of
> as a complex of intimately correlated and integrated individual
> schemas; some of them seen to predominate over the rest." (p. 500).

Gerstman stresses that the physiological and psychological
dispositions of the body image are largely unconscious although they,
may to a greater or lesser extent be reflected into or permeate into
the sphere of conscious cognition.

A recurrent problem regarding the body image is the relative
importance of central versus peripheral physiological factors. The
well known phantom limb phenomenon has been interpreted as an indication
of the predominant importance of central as compared to peripheral neural
processes in the development and in the maintenance of self-image
properties. The phantom may develop in persons who have suffered a
sudden loss of an arm or leg or parts of a limb through trauma, amputation
and the like. The phantom, if it develops, is usually particularly
vivid immediately after the loss but as the time passes the phantom limb
gradually changes and becomes shorter and smaller and usually shows a
longer persistence of the hand or the foot than of the arm or the leg.
The phantom limb phenomenon points to an unconscious inherent drive

toward maintenance of body-integrity, toward the existence of central mechanisms of stability, but the fact that phantom limbs do not always develop (never after the amputation of a limb in early childhood for instance) and never persist indefinitely indicate that although central mechanisms of stability probably exist, the body image is never exclusively centrally determined.

The importance of tactile-kinesthetic sensations for the maintenance of an integrated body image is brought out in patients with left cerebral hemiplegia. Such patients have been described as sometimes experiencing their arm or leg as estranged and separated from their own body. In the literature one finds references to hemiplegic patients who violently deny that their paralyzed arm or leg are theirs and when shown their arm or leg as being connected with their body, state: "But my eyes and my feelings don't agree, and I must believe my feelings. I know they look like mine, but I can feel they are not, and I can't believe my eyes."

Although the attitudes of hemiplegic patients toward their paralyzed limbs do not give any conclusive evidence regarding the importance of proprioceptive stimulations for the maintenance of the body schema, they do indicate that a certain disassociation of the body schema into its tactile-kinesthetic and optic components may occur, and that in cases of tactile-kinesthetic imperception of one side of the body, the undisturbed optical function may be insufficient to preserve the unity of the body image, in short, it cautions us against over-evaluating the optic factor in the formation of the body image.

Mason in his recent book "Internal Perception and Bodily Functioning" (1961) repeatedly calls attention to the fact that an individual's body image may be based upon two quite different types of perceptual processes: upon an immediate non-cognitive (emotional) inner experience on the one hand, and on a cognitive-ideational awareness on the other. Characteristic of a mentally healthy person is his ability to focus awareness not only on ideational activity and on his external environment, but also on his non-cognitive internal environment, and his ability to do so without associated cognitive conflicts, Mason maintains. In the neurotic, as in the schizophrenic and the character-disorder personality types, there appears to be a relatively lack of awareness of the non-cognitive aspect of internal perception, whether it is constituted by internal sensation,

feeling or anything else. The frequent somatic delusions described by
schizophrenics are probably more truly cognitive than sensory.

That mental patients frequently do have a fairly remote and
distant experiential relationship to their own body has been emphasized
by a number of scholars. "The body 'image' is the nucleus of the ego",
Fenichel (1945) states. And elaborating on his statement he maintains
that body image distortions usually are the earliest heralds of schizophrenic regressions. However, not only in schizophrenia, but in severe
mental illness generally, do we commonly find such distortions, i.e.
feelings of depersonalization or alienation of body parts or the whole
body, hypochondriac complaints and the reporting about bizarre body
changes. Preoccupation with body functions and the reporting of perceived changes in body structure, frequently serve as testimony to the
shifts taking place in the body image as mental illness develops.

Depersonalization has been described as an integral part of
schizophrenic symptomatology, but it has also been reported in borderline
schizophrenics and very seriously ill neurotics. For instance,
Schilder (1951) refers to depersonalization as a neurotic symptom, stating:
"Almost every neurosis has in some phase of its development symptoms
of depersonalization." Mayer-Gross (1935) suggests that depersonalization
in some degree is involved in all psychopathologies, and the same viewpoint is found in Machover's (1949) and Reich's (1942) theorizing.

However, also among presumably normal individuals do we often
find a high degree of remoteness and distance in body awareness.
Commenting upon his experiments on self-awareness among university-
students, Hefferline (1958) states:

"A person neither dead nor in flaccid paralysis presumably
should be able to discriminate the tonic condition of all or
of any part of his skeletal musculature. The first report of
a subject is likely to be that he can do this. He can feel,
he says, every part of his body. When further inquiry is made,
it often turns out that what he took to be proprioceptive discrimination of a particular body part was, however, actually a
visualization of the part or a verbal statement of its location.
Or else, to discriminate the part, he may have had to amplify
proprioception by making actual movements. With further work,
if he can be persuaded to continue, the subject may report
certain parts of his body to be proprioceptively missing.
Suppose it is his neck. He may discriminate a mass in his
head and a mass in his trunk, with what feels like simply some

empty space between. At this stage the subject is apt to remember more important things to do and his participation in the silly business ends." (p. 748).

However, some subjects may be made curious of their blank spots and continue to pay close attention to the body parts in question, Hefferline contiues. The blank spot may then suddenly become the locus of sharp pain or electric sensations. It soon becomes imperative to relax the tension, but the subject doesn't know how to do so. The situation becomes increasingly uncomfortable. If the subject is motivated to go on, and is helped with his breathing, the static tension (of cramped muscles) will frequently suddenly give way to motoric actions and outbursts which may give instant and considerable temporary relief.

Hefferline's notion that sustained attention to a body part may release muscular pain and electrical sensations closely corresponds to Braatöy's (1954) observations:

"Releasing the postural strain can sometimes...even release emotional storms. Instead of emotional thunder one may observe 'lightning' or what patients themselves describe as something resembling electric shock. One sees sudden isolated clonic seizures comprising part of the body or the whole body. Sometimes such a fit is so violent that the patient gets frightened and cautious in relation to the treatment. My hypothesis is that the sudden shock, the clonic fit, signalizes that the organism is trying to express two different reactions at the same time." (p. 184).

Hefferlines' method of asking his subjects to give an experiential report of their body awareness is a time consuming method and a method requiring high competence and skill in evaluating and interpreting the informations obtained. A much more "objective" and easy-to-handle approach is Machover's Draw-A-Person Test. The test has many advantages. It is simple to administer, it requires no material beyond a paper and a pencil, it may be completed within a fairly brief period of time, and it can be interpreted directly from the figures that are drown without any intermediary recording or scoring.

Figure drawing as an expression of bodily postures.

Assuming that a subject in drawing a person would project his body image, and further more, that a subjects' body image expresses his underlying postural configurations, we should expect the interpretations suggested in the area of figure drawings to parallel postural interpretations and findings related to postural dynamics.

In line with our previous discussion of Deutsch's psychoanalytic observations of a certain relationship between leg postures and sexual attitudes, between arm postures and self-assertive attitudes, and Malmo's experimental findings to the same effect, it is of great interest to note that the very same relationships have in fact, been specifically stressed by Machover as a principle for the interpretation of figure drawings. Thus, Machover (1951) writes:

"When males emphasize hips on the self-sex drawing, homosexual problems are implied. The crotch in the male drawing is another area where sex attitudes may be expressed. Legs, because of their proximity to genital areas, also have sexual connotations...As an extensive and protruding organ, the foot tends to have some sexual connotations...Analysis of fetishisms, common anecdotes, and introspective meanings given by tested subjects, testify to the sexual symbolism of the feet.

...most involved in contact with persons, with our own bodies are the arms and the hands...In the treatment of arms and hands, we may get information about such personality components as aspiration, confidence, efficiency, aggressiveness, or perhaps, guilt regarding interpersonal and sexual activity."(p. 358)

In a systematic scoring manual for figure-drawing items, Witkin et al. (1954) suggest very much the same interpretations; short arms, they state, reflect frustration or lack of ambition; arms out express need for contact; arms at side, express inactivity; hands exposed suggest lack of reserve in contact; a tight stance suggests withdrawal, reserve, or defensiveness in the sexual area; and reinforcements, erasures, or shading around the legs, indicate sexual consciousness or conflict.

Of interest is also their suggestion that "a high level of interest in and respect for the body are implied in good drawing skill", that "inconsistency indicates poor integration, difficulty in maintaining integration, and impulsiveness", that "large head indicates excessive emphasis and reliance on social, ideational, and control functions, with corresponding underemphasis on body and impulse life", that "stress on boundaries and margins like cuffs on shirtsleeves or trousers, indicate active control", that "a long neck reflects repressive overcontrol, and absence of neck, looseness of control", that "beads, necklace, or similar ornaments around the neck indicates rationalized control", and finally that "heavy waistline emphasis is indicative of repression in the sexual sphere".

We are mentioning all these interpretative suggestions not in order to claim their unconditional validity but to show that a number of principles of interpretation have emerged in the area of figure drawing, in the area of perception, which corresponds fairly closely to postural observations obtained in very different settings. In short, we may suggest that in drawing a person an individual primarily projects not his psychodynamic tensions and conflicts but how these tensions and conflicts are manifested in his own postural system. If an individual shows inconsistency in his drawing of different body parts we would expect this to be paralleled by postural inconsistencies; if he shows difficulty or distortion in drawing a certain body part we would expect this to be paralleled by disturbed proprioceptive feedback and awareness of the body part in question; if he shows a strong need to reemphasize for instance the neckline or the waistline, we would expect this to be paralleled by dystonicity of the neck muscles or of the muscles at the waistline (e.g. the diaphragm) respectively. But in the same way as situationally induced postures may make depth-interpretations difficult or impossible so may the same be true as regards situational (and conscious) factors in the field of figure drawings. Also in the latter field we are probably confronted with the question of layers and the difficulty introduced by tensional displacements.

However, these are minor questions as compared to the fact that postural dynamics proper seem to offer an important rationale for the interpretation of figure drawing items.

The relationship between postural tension and pain.

Returning to Hefferline's and Braatöy's observations concerning the introspective reports of subjects in a situation calling for a sustained attention to body parts, we recall that muscular pain was particularly emphasized. For some people muscular pain is their most frequent reminder of their somatic self. Thus, an understanding of muscular pain might help us to gain insight into body awareness generally.

As a starting point, let us mention Holmes and Wolff's (1950) hypothesis that one of the most common mechanisms behind muscular pain includes: 1) a relatively intense and sustained skeletal muscle contraction (the contraction producing a metabolite capable of stimulating pain endings of the muscles), and 2) a constricted vascular condition in the muscle which engenders the gradual accumulation of the noxious metabolites to a concentration sufficient to exceed the pain threshold.

The relationship between muscular pain and increased muscular tension has been demonstrated in several of Malmo's studies. E.g. headache has been shown to occur in psychiatric interviews after the appearance of increasing muscle potentials from the forehead or the neck.

How does this increasing muscle tension come about? Two explanatory principles have been offered by Malmo and his collaborators. The first one, we may call "the specific reaction hypothesis", the second one, "the conflict hypothesis". In what follows, we want to describe each of these hypotheses.

The specific reaction hypothesis states that a high level EMG potentials in a certain muscle group represents the peripheral component of a neural mechanism which is invarably mobilized under every type of stress.

Comparing mental patients with head complaints (headache, head tightness, etc.) and heart complaints (palpitations) in an experimental stress situation, Malmo et al. (1950) found that muscle potential scores from the neck region were significantly higher for the group of patients with head complaints, while the mean heart rate, heart rate variability and respiratory variability, were significantly higher for the group of patients with cardiovascular complaints. It is

concluded that pyschiatric patients with somatic complaints tend to
manifest increased responsivness in the related physiological system
upon exposure to stress generally, although the increased physiological
activation will not always give rise to the subjective symptoms at
the time of stress.

In a more recent article by Malmo (1959) it is shown how a
loud distracting noise during a tracking experiment in the case of a
42-year-old woman with complaints of muscular discomfort localized in
the left thigh, resulted in a disproportionate tension increase in the
left thigh only. Malmo also reports that in the case of a 28-year-old
woman, with complaints of a distressing feeling of tightness on the
right side of the neck, a disproportional activation was found in the
musole in the symptom area. In a third case with complaints of tension
on the left side of the neck, the sight of the EMG recording room for
the first time was itself sufficient to increase the amplitude of
muscle potentials from the symptom area so that they become appreciably
higher than those of the opposite side of the neck. In a fourth case,
a woman with complaints of tensional discomfort in the left shoulder,
the playback of a recorded interview resulted in greater tension in the
left as compared to the right shoulder, when the topic concerning her
dead sister commenced to come over the loudspekaer. Malmo (1959)
summarizes his observations as follows:

> "As far as could be determined, the EMG data from all these
> patients were consistent in suggesting that for skeletal-
> muscle tension in patients with well developed tensional
> symptoms, increasing the activation level up to a certain point
> has the effect of raising muscle tension in one localized
> muscle group, the one in which the patient complained of
> tensional discomfort. It was not necessary for the patient to
> actually feel the discomfort during the experimental session
> for this differential result to appear." (p. 383).

The fact that subjects seem to react to unspecific stress
with a disproportionate increase of tension in localized muscle groups
raises a couple of questions. Firstly, why does one individual react
to unspecific stress with increased leg tension, another with increased neck tension? Secondly, does a relationship exist between
various types of stress and the activation of a one and the same muscle
group?

These latter questions bring us over to the conflict hypothesis.
The conflict hypothesis states that high level EMG's from a muscle
group represent evidence of a state of motoric conflict involving the
muscle group in question. In one of their studies, Malmo et al. (1954)
present results favoring such a hypothesis. When the subjects were
confronted with a situation involving conflict between arm extension
and withdrawal, arm tension was found to increase the most in the very
same subjects who showed the greatest head tension when confronted with
a situation involving a conflict between approach and withdrawal of the
head. If the specific reaction hypothesis had been valid we should
have expected the subjects showing the greatest tension increase in one
area to show relatively smaller tension increase in other areas. The
fact that the same subjects showed the highest increase in both areas
may indicate that the specific reaction hypothesis don't hold for
tension-increase within different muscular areas, but only for muscular
tension changes generally as compared to other types of physiological
responses, e.g. vasomotor, respiratory, and cardiac responses. Before
digging into these matters, let us briefly mention some hypothesis
concerning the dynamics behind various types of muscular pain, i.e.
why some people develop headache, others neckache, armache, legache
and backache.

We would like to start out with what has been called the
backache syndrome, i.e. pain and tenderness arising from muscles in the
posterior aspect of the torso and associated with varying degrees of
disability. On the basis of a study of 65 patients showing this syndrome,
Holmes and Wolff (1950) suggest that they are all characterized by
a certain personality profile. They state:

> "...in patients exhibiting the backache syndrome the participation
> of body musculature in behavior is often characterized by intense
> and sustained generalized hyperfunction...in general, the patients
> exhibiting the backache syndrome are active, anxious, restless
> people who tolerate idleness and inactivity poorly...they are
> quick to react to life situations which threaten their security
> with intense feelings of resentment, frustration,...They are often
> unable to give expression to their hostile feelings. Rather,
> they 'keep them in' and ruminate for days over their difficulties
> without being able to take decisive action to remove the threat
> to their integrity and resolve the conflict...'Action' is of
> singular importance among the constellations of security props
> utilized by these patients in their social and interpersonal

behavior. Early in childhood they learn that 'doing things
well' and 'accomplishing things' wins approval, approbation and
relative immunity from criticism. Performance itself often comes
to be equated with feelings of security...They are often neat,
meticulous and perfectionistic, and usually apply themselves
assiduously to the execution of any task they undertake, regardless of whether er not it is productive of satisfaction. Their
action patterns are designed to win approval and support by
'doing for others,' 'trying to please others,' 'keeping the peace,'
and 'carrying heavy burdens of responsibility without complaint.'
When their performance is rewarded with praise or approval it is
productive of some degree of tranquility. When such approbation
is withheld or criticism given instead these patients are quick
to take offense and feel 'taken advantage of' and 'imposed upon.'
(p. 757).

The genesis of the backache syndrome, according to Holmes and
Wolff, is to be found partly in the "on guard" pattern, the constant
state of readiness of the skeletal musculature to participate in im-
pending action, and partly in situational stress, that is, in stress
arising from situations where actions prove ineffective, where actions
are blocked, or where the person is unable to make a decision on what
course of action to take. In such situations they are "held on the
spot", constantly on guard and ready to take action, but unable either
to fight or run away. It is in such a setting that the sustained
tension of the skeletal muscles which accompanies the "on guard" pattern
of behavior may become productive of discomfort and backahe, Holmes and
Wolff maintain.

In a number of case studies using EMG recordings from the
trapezius, lumbar sacrospinalis, and the hamstring muscles they demonstrate how patients with backache complaints show generalized
participation of non-essential muscles (i.e. back muscles) when asked
to perform a simple test of motor functioning (e.g. to squeeze with
right hand the examiner's fingers), and large increase in back muscles
tension when during psychiatric interviews they are lead to discuss
events touching upon their underlying conflict between action and
"turning in a good performance".

Reich (1942) too has payed attention to back pain, although
he has mostly been concerned with pain sensations from the neck and
the musculature between the shoulder-blades. Tension in the former
area, Reich equated with a "stiffnecked" and "headstrong" psychological
attitude, and tension in the latter place, with attempts of "holding

back", psychologically and literally speaking. He further suggests
that pain in the chest muscles derives from a chronic suppression of
"intolerable longings", "heartbreaking crying", "disappointment", and
"giving surrender", and pain in the superficial and deep adductor
muscles of the thighs, and in the flexor muscles of the knee (the
muscles running from the lower surface of the pelvis to the upper end
of the tibia), from suppression of genital excitations and sensations
in the pelvic floor.

Turning to headache, we recall that headache has been found
in an experimental setting to the preceded by increased tension in
the frontalis muscles. However a most bewildering problem emerges at
this point. Granted we are confronted with a motor conflict, what
sort of conflicts (in terms of approach and avoidance) can be expressed
in the frontalis muscles? As previously told, Malmo et al. (1954)
found forehead tension to characterize psychoneurotic generally, whether
headache-prone or not. In interpreting their findings, the fact that
the frontalis muscles cannot move a bony lever, and hence cannot be
directly involved in an approach-avoidance conflict, is specifically
emphasized. It may be, they state, that the high forehead tension is
a secondary phenomenon, which is preceded by tension increase in other
muscles, like those of the neck. They add, however, that so far
no reliable correlations have been found between neck-muscle potentials
and any particular stimulating conditions in terms of thematic content.

From psychoanalytic quarters the hypothesis has been launched
that headache principally stems from "super-ego aggression", thus that
it is, an expression of an intrapsychic conflict.

It is interesting to note that Malmo et al.(1950) in one of
their case studies found frontalis tension to be considerably increased
when the patient was forced to take full responsibility for the development of the interview. In another patient studied, this patient too
showing headache to arise in association with distressing feelings during
the interview, Shagass and Malmo (1954) specifically point out that
complaint of headache occurred "after narration of a traumatic event
which had overpowered her, about which she felt guilt, and from which
there was actually no escape" (p. 306). In a third patient studied
(Malmo et al. 1950), a considerable drop in frontalis muscle potentials
was found just prior to the patient stating his refusal to go on dis-

cussing a distressing topic, the patient's inner decision to leave
the field, so to speak, resulting in an instant drop in frontalis
tension.

Some observations by Reich (1942) fits well into the same
picture. However, Reich goes one step further and hypothesizes
the involvement of more specific dynamic relationships. He states:

> "Violent headaches are a very common symptom. Their localization
> is very often just above the neck, above the eyes, or in the
> forehead....How do they come about?....occipital headaches are
> due to a hypertonus in the musculature of the neck. This
> muscular attitude expresses a continuous fear that something
> dangerous may happen from the rear, being beaten in the head,
> etc. The frontal headache above the eyebrows which is felt as
> "a band around the head" is the result of a chronic raising of
> the eyebrows....This attitude (the tensning og the whole mus-
> culature of the forehead as well as that of the scalp) ex-
> presses chronic anxious anticipation in the eyes. Fully
> developed, this expression would correspond to the opening
> wide of the eyes which is characteristic of fright...."
> (p. 270).

Thus, Reich considers frontal headache as a secondary symptom
following a postural defense against fright. Postural defenses against
fright may take various forms, Reich maintains, and he continues:

> "There are patients with a facial expression which one might
> call 'haughty'. When its expression is dissolved, it turns
> out to be a defense against the expression of a frightened
> or anxious attentiveness in the face. Other patients present
> the forehead of a 'serious thinker'...This attitude has usually
> developed as a defense against anxiety...the frightened facial
> expression was turned into the 'thoughtful attitude'. Again,
> in other cases the forehead looks 'smooth', 'flat', or
> 'expressionless.' The fear of being hit on the head is always
> behind this expression." (p. 271).

The quotation shows that Reich considers forehead tension
closely linked to anticipatory processes, particularly to anxiety
and fright. One may ask, anxiety and fright about what? In a case
discussion Reich suggests that the facial attitude of anxious
attentiveness has its psyhic counterpart in defensive attitudes of
"keeping one's head", of "never to lose one's head", of "being on
guard", etc., and that these attitudes are derived from identification
with parental figures and directed toward the watching closely over

desires and wishes which have come to be "dangerous" and "threatening"
for the maintenance of the individual's equilibrium. In the case
referred to, Reich states:

> "Historically, the severe expression of eyes and forehead
> derived from an identification with her father who was a very
> moral person with a strict ascetic attitude. At a very early
> age, her father had again and again impressed on her the danger
> of giving in to sexual desires...Thus, the forehead had taken
> the place of the father in guarding against the temptation of
> giving in to a sexual desire." (p.316).

Thus, it becomes apparent that Reich considers forehead tension
very much as an expression of intrapsychic conflicts, as a second-
order defense towards conflicts primarily expressed in the motoric
system elsewhere in the body. In this respect Reich is in full agreement with Malmo's suggestion referred to. Whether one wants to call
these anticipatory and "watching over" processes superego functions or
not, is of course to a large extent a matter of definition and pre-
ferences as regards one's conceptual framework generally. The hypothesis of a positive association between super-ego aggression and
frontalis tension is a very challenging one indeed, although so far, as
far as we know, no systematic study has been undertaken in order to
test this relationship.

Summarizing our answers to the question what determines which
muscle group gets involved in symptom formation (subjective complaints
of pain), we may suggest that this is to a large extent a matter of
genetic factors, these factors shaping an indivudual's habitual or
characteristic pattern of muscular tensions (postures), and thereby
also what sort of events represent situational stress, that is, are
dangerous and threatening, and secondarily what sort of responses are
readily available to deal with such situations. For instance, in the
case of the backache syndrome the genetic root probably has an anal
nucleus, the resulting postural-defense-structure predisposing the
individual repetitiously to achieve, please, and perform, but also to
be specifically susceptible to severe threats and stress in situations,
preventing these actions, and in such situations to be "held on the
spot", eventually resulting in severe backache. Looked at from such an
angle various types of stress situations--externally considered--may have

very much the same effect on the individual subject, all directly
or indirectly touching on the same underlying conflict and thereby
activating the very same muscle groups. Consequently, a non-specific
stress situation may trigger off tension increase in various muscle
groups in different subjects depending upon their individual character
structure, that is, their individual developmental history.

However, as previously noted, far from all people do develop
increased muscle tension in stress situations, specific or non-specific.
Why do some people develop pains in their skeletal muscles, while others
show symptom formation involving the heart, the intestines and other
internal organs?

In passing it should be mentioned that symptom formations in
internal organs not necessarily exclude the possibility of muscular
tensions here too playing a predominant role. For instance, dyspnea
and precordial pain has been considered frequently caused by a sustained
contraction of the diaphragm muscles (Wolf, 1947). Furthermore, it has
been claimed (Lewis, 1954) that a central causal factor in many patients
with periodic feelings of panicky, suffocation, anxiety and alarming
heart sensations, with initial diagnosis including cardiovascular
disturbances, endocrine disorders, gastrointestinal upset, is hyperventilation. Finally it has been experimentally demonstrated that
air-swallowing tendencies under apprehension may result in air-bubbles
in the stomach, producing alarming heart sensations by pushing the
diaphragm upward. In all these cases the tonicity of skeletal muscles
may be considered of importance, although we certainly cannot explain
away the fact that in some people the exterior of the body, in others,
the interior of the body gets involved in symptom formation.

This latter point brings us over to a facinating series of
hypotheses and some fine empirical studies by Fisher and Cleveland
(1958).

Fisher and Cleveland's findings and theoretical approach.

The starting point in Fisher and Cleveland's theorizing about
body image and personality was their empirical finding that patients
with psychosomatic symptoms in the exterior body layers (arthritis,
neurodermatitis, etc.) tend to conceive implicity of their bodies as

sheathed by an armored barrier, while patients with symptoms involving
the body interior (stomach ulcers, ulcerative colitis etc.) tend to
perceive their bodies as having fluid, thin, vague, and easily
penetrated boundaries. In a sense, their finding supports the specific
reaction hypothesis. If a person experiences the boundary characteristics
of his body image as thick and impermeable, he will, if symptoms
develop - develop these symptoms preferably in the body exterior, that
is, in the skin, the striate musculature, and the vascular component
of these two systems. On the other hand. if a person experiences the
boundary characteristics of his body image as thin and easily penetrated, he will develop symptoms preferably in the body interior, in
the internal viscera, the heart, the intestines, etc.

To get at how an indivudual expreiences the boundary characteristics of his body image, a special Barrier score (and a special
Penetration score) is derived from his Rorschach responses. The Barrier
score is said to be high if his Rorschach responses emphasize the
containing, protective, and boundary-defining qualities of the periphery
of percepts, like "cave with rocky walls", "flower pot", "knight in
armor", "turtle with a shell", "something with a wall around it", etc.
It is hypothesized that the frequency of such Rorschach percepts does
correspond to an individuals basic concepts and feelings about his
own body structure. Such a hypothesis might sound rather far-fetched.
However, in a couple of psychophysiological experiments, it is demonstrated that high Barrier subjects manifest a relatively high reactivity pattern of responses involving the outer body layers and a
relatively low reactivity pattern of responses involving the body
interior, whereas the reactivity patterns are the exact opposite among
low Barrier subjects. Studying a high and a low Barrier group under
resting conditions, it was found that Barrier scores correlated
positively with EMG's (over the frontalis muscles of the left eye) and
GSR reactivity, but negatively with heart rate and systolic blood
pressure. Specifically high correlations were found between Barrier
scores and physiological reactivity changes from rest to stress. The
high Barrier subjects showed a relatively greater increment of reactivity of their exterior body layers, that is, their EMG's, GSR's,
and peripheral resistance, increased relatively more than what was the
case among the low Barrier subjects, while the latter group significant-

ly exceeded the former in the degree of increase of heart rate, stroke
volume, and cardiac output. Consequently, it is suggested that the
more definite the individual's body image boundary the greater the
increment in body exterior responses under stress, and the less increment under stress in body interior responses. Most responsive to
stress in the high Barrier group was the EMG, while this measure was
the least responsive in the low Barrier group. Among the most responsive
measure in this latter group was heart rate.

Commenting upon their empirical findings Fisher and Cleveland
state:

> "It is our supposition that if rest-level conditions could be
> sufficiently standardized for a group of subjects, the body-
> exterior versus body-interior patterns would emerge with a
> clarity as great as that which we have found to hold true for
> the rest-level stress-level difference score. We consider that
> individuals manifest habitual ratios of body-exterior to body-
> interior reactivity in their everyday living...The results of
> the present studies indicate that such habitual individual
> differences persist even at high states of emotional arousal."
> (p. 336).

In another study in the same research program Fisher and
Cleveland demonstrate that a link seems to exist between body image
boundary definiteness and EEG waves, high Barrier subjects showing less
per cent time EEG aplha waves than low barrier subjects. This fits well
with their conception of high Barrier subjects habitually being more
alert, more actively dedicated to being self-steering, more mobilized
for high aspirations, in short, more chronically activated.

In explaining their findings, Fisher and Cleveland suggest
that a relationship exists between EEG alpha waves and striate muscle
tension. They postulate that the greater the muscular tension, the more
intense is the resulting proprioceptive stimulation evoked, and the more
intense this stimulation, the more will blocking of alpha waves take
place. Behind the postulate are two assumptions. The first one, that
the blocking of alpha waves (cortical arousal) is a function of
stimulation of the brain stem reticular formation. The second one that
proprioceptive impulses are specifically potent in stimulating the
ascending reticular formation. They note that there are indications
that of all the various kinds of sensory impulses that may activate the

reticular formation, those of a proprioceptive type rank only second
to pain in potency. They quote Gellhorn (1953) to the effect that:
"It seems that afferent impulses originating in the muscle spindles
and quantitatively related to the intensity of the muscle tone determine
to a great extent the excitability of the nervous system and thereby
the state of wakefulness." (p. 186). And they add that visceral
sensations below the pain threshold apparently have much less activating
influence upon the reticular formation than do proprioceptive sensations.

We mentioned above that Fisher and Cleveland look upon a high
Barrier subject as habitually more alert, more self-steering, more
mobilized for high aspirations than a low Barrier subject. In so doing Fisher and Cleveland base their opinion upon a series of empirical
findings. Having developed their Rorschach Barrier score they contacted
a large number of other investigators who had previously made use of
the Rorschach test in the exploration of various aspects of personality
functioning, i.e. by correlating their Rorschach findings with other
more objective data independently obtained.

Comparing normal subjects with high and low Barrier scores
it emerged that high Barrier subjects show a greater tendency to express aggression outwardly when frustrated, to be able to tolerate
stress, to complete interrupted tasks, to set higher goals for themselves, to be more oriented toward self-expressiveness, to show less
suggestibility, to show more initiative, more persistence, higher
achievement orientation, in short, more self-steering behavior.

Furthermore, based upon observation of the behavior of high
and low Barrier subjects in group situations it is suggested that persons
with low Barrier score have a greater tendency to structure the groups
in such a way that one person is taking the lead and relieving everyone
else of responsibility, and to embed group procedures in a context
emphasizing formality and precise definition of role relationships.
The low Barrier person tend to compensate for his inadequate boundary
by stabilizing the demands that can be made upon him from the outside.
The more the world can be defined in advance and the more people do
things according to detailed formal rules the more the immediate
environment becomes a stabilized boundary region. A person with a weak
body image boundary will in other words seek ways to supplement his
individual boundaries by creating exterior conditions which might

artifically provide a substitute boundary capable of supplying a minimum
of constancy and security in relation to new and changing situations.
A low Barrier person will prefer certain types of group structures.
His orientation towards life will be quantitative rather than qualitative, theoretical rather than action oriented, focused on things
rather than on human beings, and oriented toward external figures for
guidance rather than on his own self-reliance.

Fisher and Cleveland hypothesize that body image boundary
definitiveness is a function of early socialization processes, that
differences exist between high and low Barrier subjects in their
childhood experiences and family background. They attempt to test this
hypothesis on two levels, 1) by comparing the Barrier scores of subjects belonging to various cultural groups with known differences in
child rearing practices, and 2) by studying and comparing the personality type of the parents to subjects with known differences in
Barrier score.

In order to study the first question a comparison is made
between the Barrier scores of subjects from the following cultural
groups: Bhil (India), Navaho and Zuni (Southwestern U.S.), Tuscarora
(Northeast U.S.), Hindu (India), Haitian, and three different U.S.
samples. Comparing the average Barrier score they arrive at a rough
division of the various cultural groups into two categories. One category consisting of the Bhil, Navaho and Zuni groups characterized by
relatively definite body image boundaries, and another category, consisting of the Haitian, a U.S. Chicago sample, and two U.S. college
samples, all characterized by significantly less definite boundaries
than the former groups. Examining the difference between the blocs of
definite and indefinite boundary cultures in terms of parental attitudes,
child rearing practices and values, they arrive at two hypotheses
accounting for the differences observed; the first one having to do
with the degree of tolerance for impulse expression, and the other,
with the degree of clearness of definitions of the standards and values
in the cultures.

Commenting upon the first hypothesis, they state:

"It appears...that the high Barrier cultures provide considerable
more freedom for the developing child to indulge his impulses
than do the low Barrier cultures. For example there seems to be
much less restraint on direct relief of body tensions in the

Bhil and Navaho groups (especially during the developmental
years) than in the general culture of the United States...
The three high Barrier groups are alike in that they all permit
relatively greater impulse satisfaction during the first few
years of life than does the modern Western mode of socialization...
the sequence which perhaps maximizes the chances of a culture
producing definite boundary individuals is one marked by an
initial period of almost complete gratification that is in turn
followed by a period in which training of adult responsibility
proceeds forcefully and on the basis of immediate realistic
issues." (pp. 283-92).

In other words, high Barrier cultures seem to permit relatively
great gratification to the child during early infancy, but tend to
make quite severe, exacting and reality-oriented demands upon the child
following the long gratifying period. That is, once the child's independence training is started, the child is pushed steadily and
aggressively. He is confronted with real life tasks and clear-cut
responsibilities are placed upon him. It strikes us, Fisher and
Cleveland state, that the modern Western culture proceeds in a manner
which represents the complete reversal of this pattern: imposition
of real life responsibilities tends to be delayed until adolescence
and continued dependence on parents into adolescence and even early
maturity, while early socialization training often is relatively severe,
demanding and punitive.

Commenting upon the second hypothesis mentioned, Fisher and
Cleveland state that is seems reasonable to say that the three high
Barrier cultures provide the individual with more definite and less
conflictual value models than does modern Western culture. The parents
in these cultures would probably be in a better position to supply a
sharply contoured model, a model possibly required for the formation
of definite body-image boundaries.

This last point is reemphasized in the discussion of the TAT
responses of parents of high and low Barrier subjects. Specifically
the mothers of high Barrier subjects were found to produce definite
stories, to show high aspirations, and to identify themselves with
forceful, well-contoured ways of behaving, tendencies to stand for
something definite and to engage themselves in intimate communication.

The same picture emerges from high and low Barrier subjects'
image of their parents as revealed in their own Rorschachs and TAT

stories. The parent image of high Barrier subjects were found to be
more definite, to be ascribed more specific attributes, to stand for
more definite values and ways of doing things. The parents of low
Barrier subjects, were conceived of as more distant, more diffuse, but
also as more threatening, destructive and disrupting. Not only was the
mode of interaction with parents apparently more poorly structured,
the parents being considered strangers, but also more often were the
parents experienced as opposed to the subjects own growth and development.

In summarizing their findings Fisher and Cleveland maintain
that a low Barrier person is more likely to grow up in a family
atmosphere characterized by instability, overtones of anxiety, in-
security about dealing with reality problems, rigidity, and chronic
tensions. His formative years are more likely to be marked by re-
strictiveness, a narrow range of permissible behavior, and frustrating
restraints which block outlets for relieving tensions. Generally, a
low Barrier person has usually grown up in a setting which is less
secure and more frustrating than that in which the high Barrier person
has developed.

Returning to their observations regarding the higher physiological reactivity in external bodily regions among high Barrier subjects, they suggest the following causal explanation:

"The reactivity of the body interior will customarily tend to
be relatively higher than that of the body exterior unless the
individual's experiences have led him to assign a certain minimum
degree of importance to the outer 'volitional' section of his
body. One is postulating that excitation tends to flow predominantly into body-interior responses until the exterior
'volitional' body layers have been stimulated through certain
socialization experiences to develop specific readiness to
respond characteristics. Consequently, conditions which do not
stimulate the 'volitional layers' would permit the body-interior
reaction-dominance to persist." (p. 313).

In another place we are told that the most meaningful body
experiences in young children probably centers around the body interior.
And that:

"It is conceivable that as the child grows older there is a
differential rate at which the (body image) boundary approaches
the body surface in various areas. At points of intensive
contact with the world (e.g. the mouth, anus, eyes) the boundary

might easily be set at the surface, but in other areas the
boundary might be conceived as lying more within the interior."
(p. 348)

One may ask if intensive contact is enough to establish body
image boundaries at the body surface? Fisher and Cleveland suggest
that also the nature of the contact might be of importance:

> "Presumably gratification of the tension arousing in the oral,
> anal, and genital areas in the different developmental phases
> leads to positive, 'good', feelings on the part of the infant
> for his body...Presumably a 'good' body feeling precedes a 'good'
> or adequate degree of ego awareness. Control of the ego or self
> must probably be first established through control of the body...
> perhaps final confidence in the integrity of one's body
> boundaries is contingent on previous satisfactory socialization
> experiences with each body area." (pp. 239-40).

In suggesting that body image boundary definiteness correspond to a high skeletal muscular tone, and that the developmental
process involved consists of a gradual reduction of a body-interior
reaction-dominance through satisfying socialization experiences, Fisher
and Cleveland deviate from our own theoretical position. Instead of
conceptualizing the development of the low Barrier subject in terms of
a fixation to an early infantile mode of reacting, we would be inclined
to look upon the process as a gradual reduction of a body-exterior
reaction-dominance through dissatisfying socialization experiences. In
viewing the process from such an angle we are in agreement with observations of children suggesting hypotonic responses to be preceded by
hypertonic ones, and we are also in agreement with the therapeutic
notion of Lowen and others that hypotonicity of external muscles expresses a fairly deep inhibition of impulses, inhibitions which have
to be transferred into hypertonicity in order to be therapeutically
accessible. We are willing to agree with Fisher and Cleveland in their
conception of a high Barrier score as an indicator of the degree to which
an individual has been able to establish a separate stabilized identity,
although we would add that a high Barrier score doesn't discriminate
between a compensatory and non-compensatory stabilized identity.

At one place Fisher and Cleveland state:

"It is possible that there is some overlap between the boundary
concept and the ego-strength concept. Perhaps ego-strength
reflects in part the degree to which the individual has evolved
boundaries definite enough to give him a clear-cut identity."
(p. 128).

We would like to add that traditionally, a number of neurotic
conditions have been considered compatible with ego-strength. In fact,
some scholars have suggested that character neuroses frequently are
characterized by too much ego-strength detrimental to the individual's
placidity, flexibility and mental health.

Our theoretical viewpoint gets some indirect support from
Fisher and Cleveland's empirical data. Comparing various nosological
groups they report that normals and neurotics show higher Barrier score
than paranoids and undifferentiated schizophrenics. The result suggest
that in the case of the psychoses a definite shift in body image occurs
and that the shifts primarily take the form of a loss and destruction of
the body image boundary, Fisher and Cleveland maintain. It is interesting to recall that Bleuler in his classical description of schizophrenia pays specific attention just to the violition of body boundaries,
the schizophrenics boundaries being obliterated or being so plastic and
vague as to be worthless either as a defense against perceived threats
or as a reference point to be used in distinguishing self from the outer
world. As regards Fisher and Cleveland's findings, what we consider
specifically important is that neurotics were found to show slightly
higher Barrier scores than normals. This last finding supports the idea
that a too rigid and definite body image boundary may have pathological
consequences, and furthermore, that high Barrier scores might correspond
to a compensatory postural rigidity as well as to a "healthy" body
awareness.

The problem of positive self-awareness.

Commenting upon their finding that neurotics obtain slightly
higher Barrier scores than normals, Fisher and Cleveland state:

"Within the range of studies we have completed, there have been
no significant indications of deficit in functioning being
associated with extremely high Barrier scores. But we made
some qualitative observations which lead us to wonder whether a

certain amount of boundary indefiniteness may not increase
one's ability to understand and play the role of others...
Presumably, a high definite boundary without the potential
for fluidity would interfere with role taking (empathic
ability)." (p. 357).

At this latter point, Fisher and Cleveland are in agreement
with a hypothesis launched by Reich in the middle thirties. In Reich's
writing about muscular or character armors, he repeatedly emphasizes
that such armors restrict an indivudual's self-awareness, his capacity
for sexual pleasure, for productive work, and for contact with his
surrounding world. However, he does not equate lack of armoring with
positive mental health. The healthy or genital character type, he
states, is able to close up in one place and open up in another. He
is in command of his armor. His armoring does not function automatically
regardless of what the external situation may be as is the case of
neurotic armoring. In his own words (1949):

"...in the case of neurotic armoring, the muscular rigidity
is chronic and automatic, while the genital character has his
armor at his disposal; he can put it in operation or out of
operation at will...It cannot be the goal of mental hygiene to
prevent the ability to form an armor; the goal can only be that
of guaranteeing the maximum vegetative mobility, in other words,
the formation of an armor which is mobile." (p. 349).

These last remarks point back to our previous discussion of
postures and perceptual behavior, but it may also serve as an introduction
to our discussion of positive self-awareness, a topic to which we will
turn in this section. This topic has been extensively discussed by
Reich. However, his theoretical analysis is so mixed up with crude
electrical speculations, his conviction that he had discovered a new
form of bio-electrical energy (orgone energy), and his assumptions so
speculative and ill-documented that most scientifically trained researchers immediately get inclined to turn their backs to his writings
and postulates. By so doing, they also, however, deprive themselves of
some of his facinating empirical observations which deserve serious
consideration and explanation.

One of Reich's observations was that parallel to the loosening
and dissolution of muscular armors, patients spontaneously reported
streaming sensations, sensations of warm pleasurable streamings going

through their bodies. Commenting upon these observations by Reich,
Braatöy (1954) states:

"These streaming sensations are among the phenomena which led
Reich into his crude electrical speculations. However, in
connection with relaxation treatment, or with direct observations on the couch in analysis, one can discover that streaming sensations go parallel to well-known clinical phenomena,
that is, to vasomotor changes. A relaxation treatment concentrated on a part of the patient's body may produce differential relaxation, accompanied by a 'differential' vasodilation
in the relaxing part of the body...The patient will register
the accompanying vasomotor changes as streaming sensations of
warmth. Like all deep sensations, and especially those connected
with the circulation system they will frequently be disturbing
and tend to produce fear, causing the patient to retreat to a
more tense, defensive, 'cold' attitude." (p. 231).

In spite of Braatöy's rejection of Reich's principle of
explanation, he is willing to follow Reich as regards the existence of
the subjective phenomenon involved, and also that the bodily sensations
reported go parallel to a sort of optimal tonicity.

What is of specific interest in the present context is Reich's
hypothesis that an individual's self-identity and self-confidence are
determined by the degree and the nature of his body-awareness. What
is of central importance, is his awareness of vegetative mobility and
wholeness, and this awareness in turn being determined by specific bodily
processes.

It is possible to interpret Reich as referring not exclusively
to a conscious awareness, but to a type of awareness that most of the
time is outside central consciousness, a sort subliminal background
stimulation which an individual carries with him unwittingly, but which
he also under certain circumstances may let permeate into his conscious
awareness. Thus, when talking about the healthy character type as
being in command of his armor, he seems to imply just such an ability
to focus awareness on postural configurations and to dissolve features
blocking "vegetative mobility".

Reich (1942) states:

"There is no doubt that the basic criterion of psychich and
vegetative health is the ability of the organism to act and
react, as a unit and as a totality, in terms of the biological
functions of tension and charge. Conversely, we have to con-

sider as pathological every non-participation of single organs
or organ-groups in the unity and totality of the vegetative
function of tension and charge, if it is chronic and represents
a lasting disturbance of the total functioning of the organism.
Clinical experience shows, furthermore, that disturbances of
self-perception really disappear only after...all organs and
organ systems of the body (are) gathered into one single experiental
unit,   with   regard to contraction as well as to expansion.
From this point of view, depersonalization becomes understandable
as a lack of charge, i.e. as disturbance of the vegetative innervation of the individual organs or organ systems, of the finger-
tips, the arms, the head, the legs, the genital, etc. Disunity
of the perception of one's own body also is caused by the interruption, in this or that part of the body, of the current of
excitation. This is particularly true of two regions; One is
the neck, where a spasm blocks the progression of the wave of
excitation from chest to head; the other is the musculature of
the pelvis, which, when spastic, interrupts the course of the
excitation from abdomen to genitals and legs. Every disturbance
of the ability to fully experience one's own body, damages self-
confidence as well as the unity of the body feeling. At the same
time, it creates the need for compensation. The perception of
one's vegetative wholeness, which is the natural and the only
safe basis for a strong self-confidence, is disturbed in all
neurotic individuals." (p. 319).

The above quotation raises a number of questions. What is
meant by perception of vegetative wholeness? Reich implies that it is
the perception of one's whole body as periodically expanding and contracting. What is meant by the vegetative function of tension and
charge? Reich implies this is the function behind the perception of
wholeness and mobility, and that it involves a rhythmical vegetative
innervation of individual organs stemming from currents of excitation
being propagated peripherally from the center of the body. Finally,
he implies that the perception of wholeness and mobility may be lost
partly by the disturbances of the vegetative innervation of an individual
organ, partly by the interruption of the current of excitation before
it reaches the organ in question (due to disturbances at an earlier,
intermediary stage).

Following Reich's train of thought so far, we would have to
ask: what is the nature of these currents of excitation, how can they
be measured, how do they manifest themselves, where do they originate?

Initially Reich showed a great deal of uncertainty about the
nature of the processes in question. In one connection (1942) he indicates that it is a matter of an oscillation between parasympthatic

and sympathetic innervations, specifically of the blood vessels, and
that these oscillations result in "a constant state of pulsation in
which the organism continues to alternate, pendelum-like, between para-
sympathetic expansion and sympathetic contraction" (p. 264). In another connection (op.cit.) he suggests that it is a matter of a conductance in terms of muscular contractions from the center to the
periphery, a rhythmical, spontaneous contraction of centrally located
muscles being transmitted through an overflow of action potentials,
from the contracting to neighbouring muscles, and thus progressively to
the total organism (p. 251). In still another connection (1949) he indicates that it is a matter of rhythmic plasmatic movements but that
essentially what moves is neither protoplasm nor muscles, but "the orgone
energy with which the body fluids are charged" (p. 359). Regrettably
Reich increasingly became a captive of this last "explanation" which
brought him further and further away from any scientific clarification
of the problem. It has to be added that both his notion of peripheral
overflow of action potentials and his conception of a general para-
sympathetic innervation of blood vessels, lacks physiological evidence,
although these explanations are much more in harmony with a scientific
outlook.

Also in his conception of the origin of the 'current of
excitation' Reich shows great uncertainty. In one connection (1942)
he suggests that the wave-like excitations originate from vegetative
center in the abdominal cavity, that is from the "large centers of
the autonomic nervous system, particularly the solar plexus, the
hypogastric plexus and the lumbosacral and pelvic plexus" (p. 260), and
that they run from the center to the periphery although they may also
run in the opposite direction, a phenomenon considered equivalent to
anxiety. In another connection, (op.cit., p. 306) he talks of the wave
of excitation running from the head downwards. In still another connection (1949, p. 373), he suggests that the waves of excitation are running
from the tail end to the front end. And finally, in still another
connection (op.cit., p. 388) that "strong waves of excitation run from
the middle of the body upwards and downwards, toward mouth and anus."
Although in the latter instance he claims that the plasmatic currents
under discussion are independent of the anatomy of muscles and nerves,
he still indicates that their point of origin in the middle of the body,

is the Plexus solaris. One may ask, why should orgonomic pulsations
outside the realm of muscles and nerves have their origin in a vegetative ganglion? How come that such pulsations outside the realm of
muscles and nerves supposedly suddenly get impeded, stopped and bound
by skeletal muscular dystonicity?

That muscular dystonicity is particularly influential in blocking currents of excitation is one of the main points in Reich's theory.
We are reminded, for instance, of Reich's statement; previously quoted,
that spasticity of two regions of the body, the neck and the pelvis,
is particularly influential in blocking currents of excitation. However, Reich, as usual, is far from consistent in his statements. In
another connection (1942) he maintains: "Muscular spasm occur with
particular predilection in parts of the organism which have annular
musculature, particularly in the throat and the anus...of similar importance is the annular musculature at the entrance to and the exit
from the stomach" (p. 313). In still another connection (1949) he
states: "...the indivudual muscular blocks do not correspond to an individual muscle or nerve. If...one looks for some rule which these
blocks must inevitably follow, one finds that the muscular armor has
a segmental arrangement...A segmental structure of the armor means that
it functions in the front, at the sides and in the back, that is, like
a ring" (p. 371). While the currents of excitation are considered always
to run along the torso, so are the blockings, in other words, seen as
rings at right angles to the body axis. Each ring or segment is defined
as comprising those organs and muscle groups which are functionally
interdependent from the point of view of expressive movement, and
differentiated from the next following segment, by those neighboring
organs and muscle groups which remain unaffected by the expressive
movements of the first segment. Reich claims that it is possible to
differentiate between seven segments in all. These are: 1) the ocular
segment, including the forehead, the eyes and the cheekbone region,
2) the oral segment including the lips, the chin, the throat and the
occipital region, 3) the neck segment, including the deep and superficial neck muscles and the root of the tongue, 4) the chest segment,
including the chest, the shoulders and the arms, 5) the diaphragmatic
segment including the diaphragm and the neighbouring organs under it,
6) the abdominal segment, including the deep and superficial abdominal

muscles and the lumbar part of the sacrospinialis and the Latissimus
dorsi, and finally, 7) the pelvic segment, including the muscles at
the pelvic floor, the thighs and the legs. Each of these segments and
potential blockings may exist, according to Reich, relatively independent of each other, although he seems to imply that the dissolution
of more peripheral segments is a prerequisite for the dissolution of
more central ones, specifically the diaphragmatic segment.

In describing the manifestation of the wave-like currents of
excitations Reich (1949) points out that they consist of relatively
slow movements. Partly he uses as a model the peristaltic movements
of an intestine, of a worm or a protozoan, and partly the rhythmical
movement of the tail end of insects, such as bees and wasps, and the
movements of a swimming jelly-fish. Here, too, he shows inconsistency.
On the one hand he compares the human organism to a living bladder
capable of peripheral contraction and expansion in all directions, and
on the other hand, to an elastic tube incapable of spherical contractions, a tube which has to bend (because of its boney structure,
i.e. the spinal column) in response to contractions. Consequently, in
this latter instance the wave-like movement considered characteristic
of organismic unity is seen only as a movement of the ends of the body
(i.e. the torso) towards and away from each other. When they are close,
there is contraction, when far apart, there is expansion and relaxation.

The conception of rhythmic, relatively slow movements of expansion and contraction involving primarily the torso, immediately calls
our attention to respiration. Are the movements and the currents of
excitation referred to by Reich synonymous with and interdependent on
respiration?

Reich's discussion of the relationship between respiration and
"currents of excitation" is rather unclear and discursive. In one
connection (1942, p. 264), discussing the basic biological functions of
expansion and contraction as an autonomic phenomenon, he states: "The
function of respiration is too complicated to be briefly presented in
terms of these new insights." In another connection (1949, p. 388),
discussing the importance of dissolving chronic diaphragmatic spams in
order to liberate the orgasm reflex, he states: "The dissolution of the
diaphragmatic block inevitably ushers in the first convultions of the
body...these convultions go with deep expiration and a wave of excitation

from the diaphragmatic region to the head on the one hand and the genitals on the other." In the third place (1942, p. 297), he states: "There is no neurotic individual who is capable of exhaling in one breath, deeply and evenly...with deep expiration, there appear in the abdomen vivid sensations of pleasure or anxiety. The function of the respiratory block (inhibition of deep expiration) is exactly that of avoiding the occurrence of these sensations...to prevent the natural vegetative rhythm of respiration... in order not to let the spontanous ve etative movement come to pass".

These two latter quotations certainly point to a close connection between currents of excitation and respiration.

Describing what he considers natural respiration, Reich (1942) states:

> "With natural respiration, the shoulders become relaxed and
> move gently and slightly forward at the end of expiration...
> The head glides back, the shoulders move forward and up, the
> middle of the abdomen draws in, the pelvis is pushed forward,
> and the legs part spontaneously." (p.298).

Again we are faced with a description rather similar to Reich's conception of "vegetative movements."

We may ask whether organismic unity in a dynamic sense is synonomous to natural respiration, whether currents of excitation are nothing else than an integral part of deep and unrestrained respiration. If we assume this to be the case, we would have no urgent need to postulate either "orgonomic pulsations" or "plasmatic contractions". All we would need would be an explanatory principle accounting for the fact that respiratory changes in the central part of the body (i.e. the torso) may produce wave-like currents of excitation which are spread outwards to peripheral parts of the body, - that these currents of excitation can be interrupted and disturbed, - and finally, that the currents in question stimulate introreceptors so that their impact is fed back into higher sensory centers influencing an individual's body awareness. Although Reich's theoretical contributions would be much more limited than what he is claiming, the model if it was found to be true would certainly represent a significant achievement in itself.

The assumption that a consistent and continuous somatic (pre-conscious) self-awareness is a central feature in an individual's ego-identity is a very plausible and reasonable one. The same is true as regards the assumption that a continuous self-awareness would require some sort of rhythmic wave-like current of excitation. It is a well known general law in physiology that all sensory mechanisms in the body are dependent upon changing fluxes of energy and that a receptor that receives a constant energy level quickly adapts itself out of operation. For instance, it is known for the carotid sinus, that the rate of change of intrasinual pressure is a much more effective stimulus than the level of pressure itself. The same principle seems to apply to baroreceptors generally and to proprioceptors and other receptors. In general, a constant and unvarying stimulus leads rapidly to sensory adaptation.

If it had only been for the two assumptions just mentioned Reich's position would have corresponded fairly closely to a theoretical model launched by Schneider (1954). Schneider suggests that the heart plays a fundamental role in an individual's enduring self-image. He points out that the heart is a somewhat unique body organ in that it produces a rhythm which is felt all over the body and that it represents a framework within which important body sensations are experienced. Finally, he also suggests that the pattern of regularity and rate shown by the heart may be correlated with central aspects of an individual's ego-integration.

The heart certainly shows a characteristic rhythm of its own, but so does also respiration. The heart produces volume changes through the pulse wave. But in this respects too do we find striking similarities with respiration.

Rhythmical changes in body volume.

Plethysmographic recordings of changes in limb volume have been a standard physiological procedure for the measurement of blood flow since the turn of the century. The changes in the volume of the ears, the fingers, the toes and the hand has been extensively studied.

One of the most systematic studies done so far, of volume changes in the fingers, toes and ears, is a study by Burch, Cohn and Neumann (1942). Not only do the investigators report continuous variations in limb volume over time in resting subjects, but they also claim that the variations observed naturally fall into five easily recognizable rhythms, one of these rhythms corresponding to the heat beats, another to the respiratory movement.

The volume change in the finger tip corresponding to each
respiratory cycle was found to be around 0.04 %. Individual differences were found in the size of these waves, variations ranging from
0.002 to 0.1 % of the finger tip volume. Also noted were changes
from time to time in the same subject.

Waves associated with the respiratory cycle have also been
observed in blood pressure recordings taken directly from the arterial
lumen. The most common conception of these waves is that they are
hydraulic in nature and caused by the changes in intrathoracic pressure
found during each respiratory cycle. In contrast to the intrapulmonic
pressure which reaches atmospheric pressure during the latter part of
each inspiratory and expiratory phase, so is the intrathoracic pressure
always negative. However, the intrathoracic pressure too changes with
inspiration and expiration. It decreases during inspiration – due to
the elastic resistance of the lungs, and increases during expiration.
If the glottis is closed an attempted expiration may increase the
pressure substantially. Changes in the intrathoracic pressure affect
the blood flow and the blood pressure in the thoracic cavity. As the
pressure increases during expiration it will influence directly the
heart and the blood vessels. It will produce a rise in the pressure of
the incompressible blood contained in these organs. The pressure on
the large arteries will be transmitted along the arterial tree to the
head and the extremities, increasing the blood pressure outside the
thoracic cavity. It is the propagation of the intrathoracic pressure
increase during expiration which is supposed to cause respiratory waves
of blood pressure.

The propagation of pressure changes can hardly be considered
currents of vegetative excitation. The existence of another type of
vascular respiratory processes is indicated by records of the impulses
reaching the arterioles however. According to Fulton (1949):

"Records of the impulses which reach the arterioles via
vasoconstriction nerves indicate that the activity of the
cells in the medullary vasoconstriction center, though continuous, is expressed by rhythmic bursts of impulses. The
rhythm of the impulse groups is often associated with the
respiratory rhythm" (p. 741).

Evans (1933) suggests that the correspondence found between oscillations of vasomotor activity and the respiratory rhythm is due to irradiation of impulses from the excited respiratory center to the vasomotor center of the medulla. Without going into physiological details it is questionable whether hydraulic respiratory waves are at all picked up plethysmographically, although they certainly may be so by intra-arterial recordings. It should also be noted that oscillations of periphereal vasomotor activity may be due to other more peripheral processes than the irradiation of impulses between medullary centers. Neurophysiological evidence shows, for instance, that pressure sensitive receptors in the carotid sinus and in the aortic arch, and probably also by other pressure-receptors in the walls of the heart itself, are extremely sensitive to even small changes of intrasinusal pressure and that these receptors participates in the cardiovascular homeostasis. Thus peripheral respiratory volume changes may not be a question of either hydraulic or vasomotor effects but interactional effects stemming from both sources. Hydraulic stimuli may produce vasomotor responses, and the latter responses in turn affect the hydraulic impact on other baroreceptors.

Granted that currents of vegetative excitations, synchronized with the respiratory rhythm, are propagated peripherally, - an assumption that is reasonable - the next question confronting us is the relationship between these impulse waves and the tonic state of skeletal muscles. The question should preferably be devided into two parts - the relationship being present centrally and the relationship being present peripherally. As regards the first part, we are dealing with the effect of the skeletal muscles involved in respiratory movements proper.

Granted that we are at least partly dealing with hydraulic waves, we should expect these waves to be stronger, the deeper and the less inhibited the respiratory excursions - and perhaps specifically, the diaphragmatic excursions. The upward and outward movement of the chest is in part the result of the diaphragms leverage against the abdominal viscera in its descent, and the forward protrusion of the abdominal wall, the result of the increased pressure in the abdominal cavity by the descent of the diaphragm. THe greater the diaphragmatic excursions, the greater would we expect the circumference changes of the

torso and also the pressure changes taking place within the intra-
thoriacic cavity.

Turning to peripheral processes we would expect the effect of
respiratory determined vascular waves to be less the higher the
genereal vasomotor tone. Does any relationship exist between this
tone and skeletal muscle tension? This is a problem that has been
extensively investigated and discussed by Jacobson (1939, 1940a, 1940b)
in an attempt to throw light on the causal factors behind essential
hypertension. On the basis of his empirical studies he concludes that
an intimate relationship seems to exist between skeletal muscle tension
and vasomotor tone. As regards the interaction between the two factors
Jacobson suggest several plausible hypotheses: 1) That skeletal muscle
contraction, even in minimal degree, is accompanied by secretion of a
pressor substance by some glands - such as the superarenal, and that
this substance is transmitted to vascular beds evoking contraction of
the muscle fibers in the vascular system. 2) That the pressor substance is not at all secreted by some central glands, but comes during
contraction from the neuromuscular junction or from the muscle tissue
itself, then passing to the blood stream evoking increased tension in
adjacent vascular beds. 3) That the mediating process is not at all
secretional, but that contracting skeletal muscle fibers stimulate
sensory end-organs in the muscle spindles, arousing afferent impulses
which proceed to excite sympathetic nerve-centers and these centers
in turn exerting pressure influence via efferent sympathetic nerves
on pheripheral blood vessels. The axon type of response would be a
special case under this hypothesis. 4) That the control of vascular
tonus is initiated by activities in the cortex or in lower brain
centers, activities closely synchronized with activities within the
somatic motor system. Thus, instead of looking upon vascular responses
as initiated by feed-back processes from skeletal contractions, the
vascular responses is seen as separate responses, as accompanying or
even preceding somatic responses, although interlocked with events
within the somatic system.

Jacobson maintains that our present knowledge is insufficient
to permit any definite choice between the hypotheses mentioned. From
our own viewpoint we would like to emphasize that except for the first

one, all the other hypotheses, make allowance for local postural tensions being interlocked with local vasomotor conditions inhibiting respiratory waves.

The very last question confronting us is related to the possibility of peripheral respiratory waves having an impact on interoceptors (i.e. baroreceptors), and the possibility of these interoceptors in turn having impact on higher sensory centers. We have already offered a provisional answer to the first part of this question in a previous section when we pointed out that pressure-sensitive receptors within the vascular system has been found to be extremely sensitive to even very small pressure changes. At this point we would like to add that no sharp division seems to exist between autonomic and somatic afferents. Afferents from autonomic receptors seem to be homologous to sensory pathways and to share central formations with the somatic system. Visceral afferents, like somatic afferents, contribute collaterals to the reticular formation and the visceral afferents collaterals to the reticular system seem to be as important a contribution to the production and maintenance of cortical arousal as somatic sensory collaterals are known to be, Lacey and Lacey (1958) maintain. They state:

> "...The fact that an intensification of sympathetic tonus is
> capable of provoking an electrical cortical activation in-
> discernible from that released by an exteroceptive stimulation
> suggests that the level of sympathetic tonus participates in
> the maintenance of a state of wakefulness in the same way as the
> incessant arrival at the brain of influxes of exteroceptive
> and proprioceptive origin" (p. 169).

In order to exemplify this point Lacey and Lacey mention a couple of studies which after distention of the carotid sinus found not only a marked fall in blood pressure but a simultaneous decrease in cortical activity. Their theoretical orientation is that an autonomic response never is an end of a chain product, but becomes a stimulus to cortical and subcortical structures, that its biological function is control and regulation of cortical and subcortical activity and that the autonomic effector system plays a large role as a mechanism of behavioral self-regulation. In a couple of empirical investigations they offer data supporting such a viewpoint.

The influence of vascular responses on behavior is suggested
by the Russian experiments on the orienting reflex. As previously
mentioned an integral part of this reflex is considered vasomotor
in nature. In their work on rhythmical volume changes in the finger
tips, the ear, and the toe tips of human subjects, Burch et al. (1942)
call attention to the fact that consistent individual differences seem
to exist in this area. While emotionally labile subjects generally show
irregular and large changes, so do senile patients and phlegmatic
subjects usually show small and regular ones. Presenting their subjects
with sudden sensory stimuli (pistol shot, cold, etc.) Burch et al.
found an abrupt downward deflection frequently to occur granted the
subject was not in a state of vasoconstriction at the time of the
stimulus - a permanent vasoconstrictor tone preventing not only spontaneous rhythmical changes but also vasomotor responses to sensory
stimuli. "The statement of patients on the tenseness of their sensations
and their appearance of being ill at ease ran parallel with the
occurence of persistent vasoconstriction", Burch et al. state. And
they add:

> "During some observations, a subject would relax and would
> respond to stimuli which were not disagreeable but would, once
> one was found to be so from a psychological point of view, react
> so that his blood vessels contracted but would not recover within
> a reasonable period of time. The vessels remained contracted.
> The patient becomes afraid and no further reactions of the blood
> vessels were detected, regardless of the stimuli applied...
> It is impossible in such a study, or in all probability in any
> other type of physiological study on the vascular system in
> conscious human beings, to overemphasize the importance of
> psychological factors." (p. 662).

Although consistent individual differences have been observed
in resting subjects, this doesn't imply that no changes take place
over time granted the immediate environment is kept constant. Once
environmental influences are stabilized, the changes observed probably
reflects what is going on in the mind of the person being examined,
Neumann et al. (1944) maintain. And they point out that records of the
rhytmical volume-changes of the finger tip seem to correspond to the
subjects' description of their emotional state at the time the records
were obtained. They think it is possible to differentiate between a
waves accompanying a fully relaxed and contented state, anxiety,
elation and resentment, depression, etc.

In closing this section we would like to mention briefly a recent investigation of Doust (1960) also focused on spontaneous autonomic oscillating systems. The main purpose of Doust's study was to ascertain whether different types of psychiatric disturbances do correspond to different rates of oscillating autonomic activities under resting conditions. Concentrating upon longitudinal observations of psychiatric patients in terms of skin temperature, capillary blood pressure, capillary blood-oxygen saturation, capillary blood flow, and a couple of other physiological indices, Doust notes that each of the measures show spontaneous oscillations of a regularly periodic sine-wave cyclic type in the resting unstimulated subject. The longest cycles were consistently found in patients with schizophrenia; the shortest in patients with epilepsy and organic brain disease. Summing up his findings Doust states that "by contrast with the practically insignificant variables of age, sex and duration of illness, psychiatric diagnosis was outstanding in differentiating between the group mean frequencies." Doust's findings fit well with the observation by others that schizophrenics have a profound inability to maintain a response set and Lacey's suggestion that "autonomic discharges may have energizing, adaptive functions which adjust receptively and reactively to the needs of the moment, and, as well, free the organism from immediate and total dependence on environmental stimulation".

It may be objected that characteristic of schizophrenics is just a total independence on environmental stimulations and an over-responsiveness to internal in contrast to external stimulations. Supposedly much more internal activity should be characteristic of schizophrenics than others. In one sense this might be true, in another sense not, depending on whether we focus on ideational activities or on the internal perception of bodily function. We are reminded at this point of Freedman et al. (1961) suggestion that the amount of illusions and hallucinatory experience produced by sensory deprivation experiments is closely related to the degree of restriction being applied on the subject's mobility and that the brain seems to produce its own self-stimulation (giving raise to hallucinatory ideational activity) when external stimulation is reduced to a minimum and kinesthetic feed-back is severely restricted. Consequently, according to this hypothesis we should expect the most disorganizing impact of a sensory deprivation

experiment, the less kinesthetic and autonomic feed-back that are produced
in a given individual.   That this is the fact is furthermore strongly
suggested by a recent sensory deprivation study by Silverman et al.
(1961) in which it was found that body-oriented subjects (subjects able
to use bodily cues in a spatial orientation test) become much less
discomforted by personal feelings and fantasies, much less inclined to
consider delusional ideas and images as having a source outside themselves, and much less time-sense distorted, than field-oriented
subjects.[1] Keeping to our theoretical model we may suppose the latter
type of subjects to be characterized by less kinesthetic feed-back,
by less autonomic fluctuations, and finally, by less respiratory
pulsations.

Some concluding remarks.

The preceding section has in many respects brought us far
behind postural dynamics and into the area of psychophysiology proper.
By referring to the results of empirical investigations and by citing
the opinions of various researchers, we have tried to show that the
Reichian model - if interpreted in a certain way - makes sense both
within a physiological and a psychological framework.   We have indicated
that strong diaphragmatic movements may very well produce wave-like
currents of excitation, expressed in the form of circumference changes
of the body, spreading upward into the chest, the neck, the head, the
arms, and the hands, and downward into the abdomen, the pelvis, the
thighs, the legs and the feet, and that the strength of excitation
reaching a certain region may not only by dependent upon its original
strength and its distance from its point of origin, but also upon the
tonicity of the muscles lying between the diaphragm and the region in
question, that is, a maximum conductability being dependent upon an
optimal level of muscular tonicity.

In formulating our theoretical position we are following
Braatöy in his conception of the streaming sensations leading Reich into
his crude electrical speculations, being explainable by vasomotor
changes.   However, we still think Reich may have hinted at something
important in emphasizing the rhythmical nature of the processes involved, the intimate relationship between somatic and vegetative

1) It should be noted that the spatial test used in a study by with-
    in et al. (1954) was found to correlate significantly with the
    Draw-a-Person Test referred to in the last chapter.

processes, and the interrelationship between central and peripheral
processes in determining an individuals self-image. In the same way
as muscular pain has been considered an interactional effect of vasomotor and skeletal muscular conditions, we may hypothesize positive
bodily awareness to show a similar basis.

Since the earliest times the diaphragm has attracted great
attention from philosophers and medical scientists alike. According
to medieval superstition the central tendon of the diaphragm is a
nervous center, the place of all emotions and the seat of the soul.
At the beginning of the nineteenth century it was demonstrated that
life could be maintained by diaphragmatic breathing alone after injuries
to the cervical spinal cord in man. What about emotions and the soul
if diaphragmatic breathing is experimentally stopped? Of course, today
with our increased knowledge of the human organism we don't expect to
murder a person psychologically by such a procedure, although our
reasoning above lead us to question if the medieval superstitions as
to the function of the diaphragm after all was completely unfounded.

Reich (1942), Fenichel (1945), Sarbin (1952) among many others,
considers an indivuduals' body self as a core element in the self-
image. If we go one step further and assume that not only is the body
image important for an individual's self-image, but also, that for the
maintenance of an individual's ego-identity, his feeling of sameness
and continuity as a human being, a continuous preconscious stream of
bodily excitations is of crucial importance, and that this stream of
excitation is generated first and foremost by the diaphragmatic excursions, we end up in a theoretical position not too distant from the
Renaissance anatomists which considered the diaphragm the _nobilissimus_
post cor musculus.

Given the assumption that respiratory pulsations influence a
person's self-awareness we would expect to find marked psychic changes
in a person following the arrest of diaphragmatic movements.

It is with great interest we learn that such an arrest has in
fact been attempted with human subjects as part of an experimental
treatment of chronic pulmonary tuberculosis. Describing the procedure
and the effects of an experiment with immobilizing lung-chamber therapy
by Barach, Kubie (1953) states:

"This (the arrest of respiratory movements) was achieved by causing increases and decreases in the density and pressure of the air in the lungs, with little or no movements either of the diaphragm or of the chest wall, for the purpose of facilitating through maximal rest the healing of the infected lung tissue. In the course of the work, (Barach) noted that a certain number of these patients, and particularily those who could reach the most complete state of respiratory rest, automatically went to some degree into a kind of hypnogogic or hypnoidal state" (p. 36).

Dealing with the relationship between respiratory arrest and changes in self-image and ego-identity, one is of course immediately reminded of the Yogis self-induced trance following specific breathing exercises. "...a kind of reverie...inevitably becomes real when relaxation follows upon pranayamic breathing". Behanan (1937) states, "Literally, pranayama means a cessation or pause in the movement of breath, i.e. prana meaning breath and ayama, pause."

The results of the experiments mentioned do not in any way necessitate or demonstrate the significance of respiratory pulsations for an individual's positive self-awareness, although they fit well into the over all theoretical framework of such a conception. Further empirical investigations in this very exciting area are very much called for. Our theoretical considerations leave in their wake a number of unanswered questions: 1) Is it possible to demonstrate empirically, circumference changes related to respiration not only in the finger tips, the ear and the toe, but also in the arm, the leg, and the head? 2) Do consistent individual differences exist in these circumference changes? 3) Are these circumference changes related not only to the depth of respiration but also to the tonicity of peripheral muscle groups? 4) Are these circumference changes related to an individuals body awareness, to his feeling of sameness and identity and self-confidence? 5) Are these circumference changes diminished in mental illness and in psychiatric patients? 6) Do stressful life events influence the changes in question? 7) What sorts of developmental factors are decisive? 8) Do we find the waves in infants? 9) Are they learned through socialization experiences? 10) May certain socialization experiences have a permanent detrimental effect? We could continue asking questions. However, the questions formulated should be sufficient to show that we are confronted with a potentially very fruitful area of research, an area spreading out into anatomy and physiology on the one hand and into social and developmental psychology on the other.

The above remarks are ment partly as a short introduction to a member of empirical investigations which are presently being planned or already under way. We don't want to go further into these studies at this point. However, we want to emphasize, that our findings so far have been rather encouraging, and that the content as well as the length of this last chapter to some extent has been determined by our own empirical observations. In a later monograph we intend to presont the results of our investigations and also to discuss further the physiological processes involved in respiratory pulsations Thus, the physiological considerations presented in previous sections are to be considered only as a sketchy and preliminary outline.

Several authors start their books by presenting synthesizing motto, frequently in the form of a quotation they consider particularity relevant as a prephase to their own contributions. Being non-verbal in orientation and consequently somewhat rebellious toward the main trend in personology, we have chosen to go in the opposite direction, namely to present our maxim at the very end:

> "...the very torm 'psychology'
> derivos from a Greek word which
> came to mean 'soul' after it
> had originally meant 'the vital
> breath'."
>
> Porls et al., 1951, p. 132.

> "...most languages demonstrate the original
> unity between breathing and psychology, re-
> spiration and spirit."
>
> Braatöy, 1954, p. 176.

REFERENCES

Abraham, K.:   Selected Papers on Psychoanalysis. New York: Basic
                      Books, 1953.

Alexander, F. & Saul, L.F.:   Respiration and personality-- a preliminary
                      report. Psychosom. Med., 1940, 2, 110-118.

Allport, G.& Vernon, P.:   Studies in Expressive Movements, New York:
                      Macmillan, 1933.

Allport, G.:   Personality. New York: Holt, 1937.

Allport, F.H.:   Social Psychology. New York: Houghton Mifflin, 1924.

Allport, F.H.:   Theories of Perception and the Concept of Structure.
                      New York: Wiley, 1955.

Ames, F.:   The hyperventilation syndrome. J. Ment. Science,
                      1955, 101, 466-525.

Ames, W.H.:   Motor correlates of infant crying. J. Genet. Psychol.,
                      1941, 59, 239-247.

Ascher, E.:   Motor attitudes and psychology. Psychosom. Med., 1949,
                      4, 228-235.

Beres, D.:   Ego deviation and the concept of schizophrenia.
                      The Psychoanalytic Study of the Child. Vol. XI.
                      New York: International Universities Press. 1956.

Behanan, K.T.:   Yoga, a Scientific Evaluation. New York: Macmillan, 1937.

Birdwhistell, R.:   Introduction to Kinesics. Louisville, Kentucky:
                      Univ. Louisville, 1952.

Blum, G.S.:   An experimental reunion of psychoanalytic theory
                      with perceptual vigilance and defense. J. Abnorm. Soc.
                      Psychol., 1954, 49, 94-98.

Blum, G.S.:   Perceptual defense revisited. J. Abnorm. Soc. Psychol.,
                      1955, 51, 24-29.

Braatöy, T.:   De Nervöse Sinn. Medicinsk Psykologi og Psykoterapi.
                      (The Nervous Temperment. Medical Psychology and
                      Psychotherapy). Oslo: Cappelen, 1947.

Braatöy, T.:   Psychology vs. anatomy in the treatment of arm
                      neuroses with physiotherapy. J. Ner. Ment. Disease,
                      1952, 115, 215-245.

Braatöy, T.:   Fundamentals of Psychoanalytic Technique. New York:
                      Wiley, 1954.

Bruner, J.S.:   On perceptual readiness. Psychol. Rev., 1957, 64,
                      123-152.

Buroh, G.E., Cohn, A.E., & Neumann, C.: A study by quantitative methods
                      of the spontaneous variations in volume of the finger
                      tip, toe tip, and posterosuperior portion of the
                      pinna in testing normal white adults. Am. J. Physiol.,
                      1942, 136, 433-447.

Burch, G.E., Cohn, A.E., & Neumann, C.: Reactivity of intact blood vessels of the fingers and toes to sensory stimuli in normal resting adults, in patients with hypertension and in senile subjects. J. Clin. Invest., 1942, 21, 655-664.

Burrow, K.T.: Science and Man's Behavior. New York: Philosophical Library, 1953.

Burrow, K.T.: Kymograph records of neuromuscular (respiratory) patterns in relation to behavior disorders. Psychosom. Med., 1941, 3, 174-186.

Christiansen, B.: Some Comments on the Causes and Effects of Parental Attitudes toward Child Rearing. Nord. Psykologi, 1961, 13, 37-116.

Christie, R.V.: Some types or respiration in neuroses. Quart. J. Med., 1935, 4, 427-432.

Clausen, J.: Respiration movement in normal, neurotic and psychotic subjects. Acta psychiat. et neurol., Suppl. 68, 1951. Pp. 74.

Clement, F.V.: Nitrous Oxide Oxygen Anesthesia. Philadelphia, Lea Febiger, 1951.

Coppen, A.F. & Mezey, A.G.: The effects of Sodium Amytal on the respiratory abnormalities of anxious patients. J. Psychosomatic Res., 1960, 5, 52-55.

Corwin, W. & Barry, N.: Respiratory plateaux in "Daydreaming" and in schizophrenia. Am. J. Psychiatry. 1940, 97, 308-318.

Darwin, C.: The Expression of Emotions in Man and Animals. New York: Appleton, 1873.

Davis, F.H. & Malmo, R.B.: Electromyographic recording during interview. Am. J. Psychiatry., 1951, 107, 908-915.

Davis, R.C.: The relation of muscle action potentials to difficulty and frustration. J. Exp. Psychol., 1938, 23, 141-158.

Davis, R.C.: Physiological responses as a means of evaluating information. In The Manipulation of Human Behavior (Eds. Biderman & Zimmer) New York: Wiley, 1961.

Deming F. & Washburn, A.H.: Respiration in infancy. I. A method of studying rates, volume and character of respiration with preliminary report of results. Am. J. Dis. Child, 1935, 49, 108-124.

Deutsch, F.: Analysis of postural behavior. Psychoanalyt. Quart., 1947, 16, 195-213.

Deutsch, F.: Analytic posturology. Psychoanalyt. Quart., 1952, 21, 196-214.

Deutsch, F.: Thus speaks the body - an analysis of postural behavior. Transact. N.Y. Academy of Sciences, Series II., 1949, 12, 58-62.

DiMascio, A., Boyd, R.W. & Greenblatt, M.: Physiological correlates of tension and antagonism during psychotherapy. Psychosom. Med., 1957, 19, 99-104.

Doust, J.W.: Spontaneous endogenous oscillating systems in autonomic and metabolic effectors; their relation to mental illness. J. Nerv. Ment. Dis., 1960, 131, 335-347.

Duffy, E.: Level of muscle tension as an aspect of personality. J. Gen. Psychol., 1946, 35, 161-171.

Duffy, E.: The psychological significance of the concept of "arousal" or "activation. Psychol. Rev., 1957, 64, 265-275.

Efron, D.: Gesture and Environment, New York: King's Crown, 1941.

Erikson, E.H.: Childhood and Society. New York, Norton, 1950.

Evans, C.L.: Starling's Principles of Human Physiology. Philadelphia: Lea & Febiger, 1933.

Faulkner, W.B.: The effect of the emotions upon diaphragmatic functions. Psychosom. Med., 1941, 3, 187-189.

Feldman, W.M.: The Principles of Anti-nasal and Post-nasal Child Physiology. New York: Longmans Green, 1920.

Feleky, A.: Influence of the emotions on respiration. J. Exper. Psychol., 1916, 1, 218.

Fenichel, O.: The Psychoanalytic Theory of Neuroses. New York: Norton, 1945.

Finesinger, J.E., & Mazick, S.G.: The effect of a painful stimulus and its recall upon respiration in psychoneurotic patients. Psychosom. Med., 1940, 2, 331-368.

Finesinger, J.E. & Mazick, S.G.: The respiratory response of psychoneurotic patients to ideal and to sensory stimuli. Am. J. Psychiatry., 1940, 97, 27-48.

Finesinger, J.E.: The spirogram in certain psychiatric disorders. Am. J. Psychiatry., 1943, 100, 159-168.

Finesinger, J.E.: The effect of pleasant and unpleasant ideas on the respiratory patterns (spirogram) in psychoneurotic patients. Am. J. Psychiatry, 1944, 100, 659-667.

Fischer, L.K. The significance of atypical postural and grasping behavior during the first year of life. Am. J. Orthopsychiatry, 1958, 28, 368-375.

Fisher, S.: Prediction of body exterior vs. body interior reactivity from a body image schema. J. Pers., 1959, 27, 56-62.

Fisher, S.: Head-body differentiations in body image and skin resistance level. J. Abnorm. Soc. Psychol., 1960, 60, 283-285.

Fisher, S.: Body image and asymmetry of body reactivity. J. Abnorm. Soc. Psychol., 1958, 57, 292-298.

Fisher, S.:   Body reactivity gradients and figure drawing variables. J. Consult. Psychol., 1959, 23, 54-59.

Fisner, S. & Abercrombie, J.: The relationship of body image distortions to body reactivity gradients. J. Pers., 1958, 36, 320-329.

Fisher, S. & Cleveland, S.E.: Body Image and Personality. Princeton, N.J.: Van Nortrand, 1958.

Fisher, S. & Cleveland, S.E.: An approach to physiological reactivity in terms of a body image schema. Psychol. Rev., 1957, 64, 26-37.

Fisher, S., & Clevaland, S.E.: Body image and style of life. J. Abnorm. Soc. Psychol., 1956, 52, 373-379.

Fisher, S. & Fisher, R.L.: Style of sexual adjustment in disturbed women and its expression in figure drawings. J. Psychol.; 1952, 34, 169-179.

Fitz, A.W.:   A study of respiratory movements. J. Exper. Med., 1896, 1, 677-692.

Freedman, S.F., Grunebaum, H.U. & Greenblatt, M.: Perceptual and cognitive changes in sensory deprivation. In Sensory Deprivation. (Ed. Solomon, P., et al.) Cambridge, Mass.: Harvard University Press, 1961.

Freeman, G.L.:   Postural tension and the conflict situation. Psycholl. Rev., 1939, 46, 226-240.

Frenkel-Brunswik, E.: Personality theory and perception. In Perception - An Approach to Personality, (eds. Blake, R.R. & Ramsey, G.W.) New York: Ronald Press, 1951.

Freud, A.:   Observations in child development. In The Psycho-analytic Study of the Child. Vol. VI. New York: International Universities Press, 1951.

Freud, S.:   Three contributions to the theory of sex. In The Basic Writings of Sigmund Freud. New York: Random House, 1938.

Fulton, J.F.:   A Textbook of Physiology. Philadelphia: W.B. Saunders, 1949.

Gellhorn, E.:   Proprioception and the motor cortex. Brain, 1949, 72, 35-62.

Gellhorn, E.:   The influence of alteration in posture of limbs on cortically induced movements. Brain, 1948, 71, 26-33.

Gerstmann, J.:   Psychological and phenomenological aspects of disorders of the body image. J. Nerv. Ment. Dis., 1958, 126, 499-512.

Gesell, A.L.:   Infant and Child in the Culture of Today. New York: Harpers, 1943.

Goldensohn, E.S.: Role of the respiratory mechanism. Psychosom. Med., 1955, 17, 377-382.

Golla, F., Haun, S.A., & Marsh, R.G.B.: The respiratory regulation in
          psychotic subjects. J. Ment. Science, 1928, 74,
          443-453.

Gray, H.:   Anatomy of the Human Body. (Ed. by C. Mayo Goss.)
          Philadelphia: Lea & Febiger, 1948.

Gray, H.:   Anatomy of the Human Body. (Ed. by W. H. Lewis.)
          Philadelphia: Lea & Febiger, 1942.

Greene, F. A. & Coggeshall, H.D.: Clinical studies of respiration:
          1. Plethysmographic study of quiet breathing and
          the influences of some ordinary activities on the
          expiratory position of the chest in man. Arch. Int.
          Med., 1933, 52, 44-56.

Gottschalk, L.A., Serota, H.M. & Shapiro, L.B.: Psychological conflict
          and neuromuscular tension. Res. Publ., Ass. Nev.
          & Ment. Dis., 1950, 29, 735-743.

Grieg, A. et al.: Nic Waal's metode for somatisk psychodiagnostikk.
          (Nic Waal's method for somatic psychodiagnostics).
          Oslo, 1957.

Haavardsholm, B.: Forsök på Elektromyografisk Registrering av den
          Nevrotiske Åndedrettssperre. (An Attempt toward
          Electromyographic Recording of the Neurotic
          Breathing Restraint). M.A. Theses. Oslo:
          Psychological Institute, 1946.

Halverson, H.M.: Variations in pulse and respiration during different
          phases of infant behavior. J. Genet. Psychol., 1941,
          59, 259-330.

Halverson, H.M.: Mechanisms of early infant feeding. J. Genet. Psychol.,
          1944, 64, 185-223.

Harris, J.B., Hoff, H.E., & Wise, R.A.: Diaphragmatic flutter as a
          manifestation of hysteria. Psychosomat. Med., 1954,
          16, 56-66.

Hefferline, R.F.: The role of proprioception in the control of behavior.
          Transact. N.Y. Academy of Sciences. Sec. II, 1958,
          20, 739-764.

Homes, T.H. & Wolff, H.G.: Live situations, emotions and backache.
          Res. Publ., Ass. Nerv. & Ment. Dis., 1950, 29,
          750-772.

Howell, W.H.: A Textbook of Physiology. Philadelphia: Saunders, 1930.

Jacobson, E.: Progressive Relaxation. Chicago: University of
          Chicago Press, 1938.

Jacobson, E.: Variations in blood pressure with skeletal muscle
          tension and relaxation. Ann. Int. Med., 1939, 12,
          1194-1212.

Jacobson, E.: Variation of blood pressure with skeletal muscle.
          tension and relaxation. II. The heart beat. Ann. Int.
          Med., 1940, 13, 1619-1625.

Jacobson, E.:   Variations of blood pressure with brief voluntary
                   muscular contractions. J. Lab. Clin. Med., 1940,
                   25, 1029-1037.

Jacobson, E.:   The effect of daily rest without training to relax
                   on muscular tonus. Am. J. Psychol., 1942, 55,
                   248-254.

Jahoda, M.:   Current Concepts of Positive Mental Health. New York:
                   Basic Books, 1958.

Jenness, A. & Wible, C.L.:   Respiration and heart action in sleep and
                   hypnosis. J. Gen. Psychol., 1937, 16, 197-222.

Katzell, R.A.:   Relations between the activity of muscles during
                   preparatory set and subsequent overt performance.
                   J. Psychol., 1948, 26, 407-436.

Kempf, E.F.:   Affective-respiratory factors in catatonia. Med. Journal
                   & Record, 1930, 131, 181-185.

Kennedy, F.:   In discussion of Finesinger and Mazick's (1940) paper.

Kitzinger, S.:   The Experience of Childbirth. London: Gollancz, 1962.

Kris, E.:   Psychoanalytic Explorations in Art. New York:
                   International Universities Press, 1952.

Kubie, L.S.:   Some implications for psychoanalysis of modern concepts
                   of the organization of the brain. Psychoanalytic
                   Quart., 1953, 22, 21-68.

Lacey, J.I., & Lacey, B.C.:   The relationship of resting autonomic
                   activity to motor impulsivity. In Brain and Human
                   Behavior. Proc. Ass. Res. Nerv. Ment. Dis., Vol. 36.
                   Baltimore: Williams & Wilkins, 1958.

Lacey, J.I.:   Psychophysiological approaches to the evaluation
                   of psychotherapeutic process and outcome. In
                   Research in Psychotherapy (Eds. E.A. Rubenstein &
                   H.B. Parloff) Washington: Am. Psychol. Ass., 1959.

Lazarus, R.S. & McCleary, R.A.:   Autonomic discrimination without
                   awareness: A study of subception. Psychol. Rev.,
                   1951, 58, 113-122.

Lewis, B.I.:   Chronic hyperventilation syndrome, J. Am. Mes. Assn.,
                   1954, 155, 1204-1208.

Lowen, A.:   Physical Dynamics of Character Structure. New York:
                   Grune & Stratton, 1958.

Lundervold, A.:   An electromyographic investigation of tense and
                   relaxed subjects. J. Nerv. & Ment. Dis., 1952, 115,
                   512-525.

McDougall, W.:   The Group Mind. Cambridge: Cambridge Univ. Press,
                   1920.

McGinnies, E.:   Emotionality and perceptual defense. Psychol. Rev.,
                   1949, 56, 244-251.

Machover, K.:   Personality Projection in the Drawing of the Human
                   Figure. Springfield, Ill.: C.C. Thomas, 1949.

Machover, K.:   Drawing of the human figure: a method of personality investigation. In An Introduction to Projective Techniques. (Eds. H.H. & C. L.Anderson) New York: Prentice-Hall, 1951.

Mahl, G.F., Danet, B. & Norton, N. Reflection of major personality characteristics in gestures and body movements. (Min. report), A.P.A. Annual meeting, 1959.

Malmo, R.B.:   Activation: A neuropsychological dimension. Psychol. Rev., 1959, 66, 367-386.

Malmo, R.B.:   Anxiety and behavioral arousal. Psychol. Rev., 1957, 64, 276-287.

Malmo, R.B., Davis, J.F. & Barza, S.: Total hysterical deafness: an experimental case study. J. Personality, 1952, 21, 188-204.

Malmo, R.B. & Shagass, C.: Physiological studies of reaction to stress in anxiety and early schizophrenia. Psychosomat. Med., 1949, 11, 9-24.

Malmo, R.B. & Shagass, C.: Physiologic study of symptom mechanisms in psychiatric patients under stress. Psychosomat. Med., 1949, 11, 25-29.

Malmo, R.B., Shagass, C., Belanger, D.J. & Smith, A.A.: Motor control in psychiatric patients under experimental stress. J. Abnorm. Soc. Psychol., 1951, 46, 539-547.

Malmo, R.B. & Shagass, C. & Davis, F.H.: Symptom specificity and body reactions during psychiatric interview. Psychosom. Med., 1950, 12, 362-376.

Malmo, R.B., Shagass, C. & Davis, F.H.: Specificity of body reactions under stress. A physiological study of somatic symptom mechanisms in psychiatric patients. Res. Publ., Ass. Nerv. & Ment. Dis., 1950, 29, 231-261.

Malmo, R.B., Smith, A.A., & Kohlmeyer, W.A.: Motor manifestation of conflict in interview: A case study. J. Abnorm. Soc. Psychol., 1956, 52, 268-271.

Malmo, R.B., Boag, T.J. & Smith, A.A.: Physiological study of personal interaction. Psychosom. Med., 1957, 19, 105-119.

Malmo, R.B. & Smith, A.A.: Forehead tension and motor irregularities in psychoneurotic patients under stress. J. Personality, 1954, 23, 391-406.

Malmo, R.B., Shagass, C. & Davis, J.F.: A method for the investigation of somatic response mechanisms in psychoneurosis. Science, 1950, 112, 325-328.

Malmo, R.B., Shagass, C. & Davis, J.F.: Electromyographic studies of muscular tension in psychiatric patients under stress. J. Clin. Exper. Psychopath., 1951, 12, 45-66.

Malmo, R.B., Wallerstein, H. & Shagass, C.: Headache proneness and mechanisms of motor conflict in psychiatric patients. J. Pers. 1953, 22, 163-187.

Mason, R.E.: <u>Internal Perception and Bodily Functioning</u>. New York: Intern. Univ. Press, 1961.

Mayer-Gross, W.: On depersonalization. <u>Brit. J. Med. Psychol.</u>, 1935, <u>15</u>, 103-109.

Meyer, D.R.: On the interaction of simultaneous responses. <u>Psychol. Bull.</u>, 1953, <u>50</u>, 204-220.

Mezey, A.G. & Coppen, A.J.: Respiratory adaptation to exercise in anxious patients. <u>Clin. Sci.</u>, 1961, <u>20</u>, 171-175.

Miles, W.R. & Behanan, K.T.: A metabolic study of three unusual learned breathing patterns practiced in the cult of Yoga. <u>Am. J. Physiol.</u>, 1934, <u>199</u>, 74-75.

Murray, H.A.: The effect of fear upon estimates of the maliciousness of other personalities. <u>J. Soc. Psychol.</u>, 1933, <u>4</u>, 310-329.

Neilson, F.M. & Roth, P.: Clinical spirography: Spirograms and their significance. <u>Arch. Int. Med.</u>, 1929, <u>43</u>, 132-138.

Neumann, C., Lhamon, W.T. & Cohn A.E.: A study of factors (emotional) responsible for changes in the pattern of spontaneous rhythmic fluctuations in the volume of the vascular bed of the finger tup. <u>J. Clin. Invest.</u>, 1944, <u>23</u>, 1-9.

Ostfeld, A.M. et al.: Studies in headache. <u>Psychosom. Med.</u>, 1957, <u>19</u> 199-208.

Paterson, A.S.: The respiratory rhythm in normal and psychiatric subjects. <u>J. Neur. Psychopath.</u>, 1935, 16, 36-53.

Paterson, A.S.: The depth and rate of respiration in normal and psychotic subjects. <u>J. Neurol. Psychopath.</u>, 1934, <u>14</u>, 323-331.

Perls, F.S. et al.: <u>Geltalt Therapy</u>. New York: Julian Press, 1951.

Piersol, G.: <u>Human Anatomy</u>. Philadelphia: Tippincott, 1907.

Plutchik, R.: The role of muscular tension in maladjustment. <u>J. Gen. Psychol.</u>, 1954, <u>50</u>, 45-62.

Rank, B. & MacNaughton, D.: A clinical contribution to early ego development. <u>In The Psychoanalytic Study of the Child</u>. Vol. V. London: Imago, 1950.

Razran, G.: The observable unconscious and inferable conscious in current Soviet psychophysiology. <u>Psychol. Rev.</u>, 1961, <u>68</u>, 81-147.

Reich, W.: <u>The Function of the Orgasm.</u> New York. Orgone Institute Press, 1942.

Reich, W.: <u>Character Analysis</u>, New York: Orgona Institute Press, 1949.

Ribble, M.A.: The significance of infantile sucking for the psychic development of the individual. In S.S. Tomkins (ed.) <u>Contemporary Psychopathology</u>. Cambridge, Mass.: Harvard Univ. Press, 1943.

Ribble, M.A.:   Infantile experience in relation to personality
                development. In F. McV. Hunt (ed.) Personality
                and the Behavior Disorders. Vol. II. New York:
                Ronald Press, 1944.

Ribble, M.A.:   The Rights of Infants. New York: Columbia Univ. Press,
                1943.

Ribble, M.A.:   Anxiety in infants and its disorganizing effects.
                In Modern Trends in Child Psychiatry (eds. N.D.C. Lewis
                and B.L. Pacelli). New York: International Univ.
                Press, 1945.

Sarbin, R.R.:   A preface to a psychological analysis of the self.
                Psychol. Rev., 1952, 59, 11-22.

Schilder, P.:   The Image and Appearance of the Human Body. New York:
                International Universities Press, 1950.

Schneider, D.E.: The image of the heart and the synergic principle
                in psychoanalysis (psychosynergy). Psychoan. Rev.,
                1954, 41, 197-215.

Shagass, C., & Malmo, R.B.: Psychodynamic themes and localized muscular
                tension during psychotherapy. Psychosom. Med., 1954,
                16, 295-314.

Shakow, D.:   Segmental set - a theory of the formal psychological
                deficit in schizophrenia. Arch. Gen. Psychiatr.,
                1962, 6, 1-17.

Simons, D.J., et al.: Experimental studies on headache: muscle of the
                scalp and neck as sources of pain. Proc. Ass., Res.
                Nerv. & Ment. Dis., 1943, 23, 228-244.

Silverman, A.F. et al.: Psychophysiological investigations in sensory
                deprivation, the body-field dimension. Psychosom. Med.,
                1961, 23, 48-62.

Smith, A.A.:   An electromyographic study of tension in interrupted
                and completed tasks. J. Exper. Psychol., 1953, 46,
                32-

Sines, F.O.:   Conflict-related stimuli as elicitors of selected
                physiological responses. J. Proj. Tech., 1957, 21,
                194-198.

Smock, C.D. & Thompson, G.G.: An inferred relationship between early
                childhood conflicts and anxiety responses in adult
                life. J. Pers., 1954, 23, 88-99.

Spitz, R.A.:   The psychogenic diseases in infancy: an attempt at
                their etiological classification. In The Psycho-
                analytic Study of the Child. Vol. VI. New York:
                International Univ. Press, 1951.

Spitz, R.A.:   Hospitalism. In The Psychoanalytic Study of the Child.
                Vol. I. New York: International Univ. Press, 1945.

Spitz, R.A.:   Hospitalism. a follow-up report. In The Psycho-
                analytic Study of the Child. Vol. II, New York:
                International Univ. Press, 1946.

Spitz, R.A.:   The smiling response: a contribution to the ontogenesis of social relations. Genet. Psychol. Monogr., 1946, 14, 57-125.

Spitz, R.A.:   Anaclitic depression. In The Psychoanalytic Study of the Child. Vol. II. London: Imago, 1947.

Spitz, R.A.:   No and Yes. On the Genesis of Human Communication. New York: International Univ. Press, 1957.

Spitz, R.A. & Wolf, K.M.:   Auteoerotism. Some empirical findings and hypotheses on three of its manifestations in the first year of life. In The Psychoanalytic Study of the Child. Vol. III-IV. New York: International Univ. Press, 1949.

Stevenson, I. & Ripley, H.S.:   Variations in respiration and in respiratory symptoms during changes in emotion. Psychosom. Med., 1952, 14, 476-490.

Surivello, W.W.:   Psychological factors in muscle action potentials: EMG gradients. J. Exper. Psychol., 1956, 52, 263-272.

Sutherland, G.I.:   The respiratory "finger print" of nervous states. Med. Rec., 1938, 148, 101-103.

Thompson, J.W., Corwin W. & Aste-Salazar, J.H.:   Physiological patterns and mental disturbances. Nature, 1937, 140, 1062-1063.

Thompson, J.W. & Corwin, W.:   Correlations between patterns of breathing and personality manifestations. Arch. Neurol. and Psychiat., 1942, 47, 265-270.

Waal, N.:   A special technique of psychotherapy with an autistic child. In The Emotional Problems of Early Childhood. (Ed. G. Caplan). New York: Basic Books, 1958.

Wade, O.L.:   Movements of the thoracic cage and diaphragm in respiration. J. Physiol., 1954, 124, 193-212.

Whatmore, G.B. & Ellis, R.M.:   Some neurophysiologic aspects of depressed states. Arch. Gen. Psychiatry, 1951, 1, 70-80.

White, R.H.:   Motivation reconsidered: The concept of competence. Psychol. Rev., 1959, 66, 297-333.

Williams, Mary F.:   Tension and respiratory pattern in young children. J. Genetic Psychol., 1942, 60, 71-84.

Wolf, S.:   Sustained contraction of the diaphragm, the mechanism of a common type of dysnea and precordial pain. J. Clin. Invest., 1947, 26, 1201.

Witkin, H.A. et al.:   Personality Through Perception. New York: Harper, 1954.

Wittkower, E.:   Further studies in the respiration of psychotic patients. J. Ment. Science, 1934, 80, 692-704.

Wolff, H.G.:   Headache and Other Head Pain. New York: Oxford, 1948.

Woodworth, R.S.:   Experimental Psychology. New York: Holt, 1938.

Supplementary references.

Feldenkrais, M.:   Body and Mature Behaviour.   London: Routledge, Kegan Paul, 1949.

Groddeck, G.:   Exploring the Unconscious.   London: C. W. Daniel, 1933.

BODY MOVEMENT
Perspectives in Research
An Arno Press Collection

Christiansen, Bjørn
Thus Speaks the Body: Attempts Toward a Personology from the Point of View of Respiration and Postures. Oslo, Norway, 1963

Dewey, Evelyn
Behavior Development in Infants: A Survey of the Literature on Prenatal and Postnatal Activity 1920–1934. New York, 1935

Evolution of Facial Expression: Two Accounts
  a. Andrew, R. J.
     The Origin and Evolution of the Calls and Facial Expressions of the Primates (Reprinted from *Behaviour,* Vol. 20, Leiden, Netherlands, 1963)
  b. Huber, Ernst
     Evolution of Facial Musculature and Facial Expression. Baltimore, 1931

Facial Expression in Children: Three Studies
  a. Washburn, Ruth Wendell
     A Study of the Smiling and Laughing of Infants in the First Year of Life (Reprinted from *Genetic Psychology Monographs,* Vol. 6, Nos. 5 & 6, Worcester, Mass., 1929) November-December, 1929
  b. Spitz, René A., with the Assistance of K. M. Wolf
     The Smiling Response: A Contribution to the Ontogenesis of Social Relations (Reprinted from *Genetic Psychology Monographs,* Vol. 34, Provincetown, Mass., 1946) August, 1946
  c. Goodenough, Florence L.
     Expression of the Emotions in a Blind-Deaf Child (Reprinted from *Journal of Abnormal and Social Psychology,* Vol. 27, Lancaster, Pa., 1932)

Research Approaches to Movement and Personality
  a. Eisenberg, Philip
     Expressive Movements Related to Feeling of Dominance (Reprinted from *Archives of Psychology,* Vol. 30, No. 211, New York, 1937) May, 1937
  b. Bartenieff, Irmgard and Martha Davis
     Effort-Shape Analysis of Movement: The Unity of Expression and Function. New York, 1965
  c. Takala, Martti
     Studies of Psychomotor Personality Tests I. Helsinki, Finland, 1953

Wolff, Charlotte
A Psychology of Gesture. Translated from the French Manuscript by Anne Tennant. 2nd edition. London, 1948